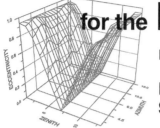

Radio Science for the Radio Amateur

by **Eric P. Nichols, KL7AJ**

Ham Radio and The Pursuit of Scientific Exploration and Discovery

Published by
ARRL 100 YEARS

Contributing Editor: Edith M. Lennon, N2ZRW

Cover Design: Sue Fagan, KB1OKW

Technical Illustration: David Pingree, N1NAS

Production: Jodi Moran, KA1JPA
Shelly Bloom, WB1ENT

Proof Readers: Judith Kutzko
Nancy Hallas, W1NCY

Cover photo: Courtesy of NASA

Copyright © 2013 by The American Radio Relay League, Inc.

Copyright secured under the Pan-American Convention

International copyright secured.

All rights reserved. No part of this work may be reproduced in any form except by written permission of the publisher. All rights of translation are reserved.

Printed in the USA

Quedan reservados todos los derechos

ISBN: 978-0-87259-338-1

First Edition

Table of Contents

Foreword

Preface — Unfinished Business

Acknowledgments

Introduction — The Radio Amateur and Scientific Investigation

Chapter 1 — On Mindset, Measurement, and Making Good Science

Chapter 2 — Safety First

Chapter 3 — The Basics

Chapter 4 — Data Acquisition — The Road from Concept to Confirmation

Chapter 5 — Software Tools for Hard Numbers

Chapter 6 — Optics — Where Radio Sees the Light

Chapter 7 — The Electromagnetic Spectrum at a Glance

Chapter 8 — Free Space Radio

Chapter 9 — A Polar Exploration, Practically Speaking

Chapter 10 — The Big Picture — Plasma as Pachyderm

Chapter 11 — The Complex Ionosphere — Weird and Wonderful Non-Linear Phenomena

Chapter 12 — Ionospheric Science in a Can

Chapter 13 — Smith Charts, Scattering Parameters, and Sundry Science Tools

Chapter 14 — Put *SPICE* in Your Life for Circuit Simulation

Chapter 15 — Antenna Modeling from the *NEC* Up

Chapter 16 — Large-Scale DAQ and Networking

Chapter 17 — Graphs and Graphics for a Big Picture

Chapter 18 — Rolling Your Own — Building Instruments for Radio Science

Chapter 19 — Now for Some Real Radio Science

Appendix I — Test Questions and Answers

Appendix II — Alphabetical Formulary

Foreword

The label "Big Science" most often describes expensive, intricate projects. But the adjective "big" may also be aptly applied to experiments distributed over large geographical areas. Prosaic needs, like power and communications, become critical concerns for such experiments. In the last decade, physicists studying cosmic rays solved these problems by employing public schools in the detection of extended air showers from high-energy particles hitting the upper atmosphere. In this book, Eric Nichols, KL7AJ, astutely observes that Amateur Radio operators may play a similar role for radio detection.

The author walks the ham radio enthusiast from fundamentals to specifics. The book begins by laying out the elementary physical science all radio amateurs should know to understand the interaction of their equipment with terrestrial phenomena. Mr. Nichols brings more than academics to the task: The message is clearly conveyed through the lens of his decades of hands-on experience in cutting-edge plasma physics laboratories. Even safety advice is delivered in the light of some unwelcome events the author encountered. Beginners will benefit from the author's experience in performing science, in particular from his skepticism toward first experimental results. Yet, Mr. Nichols also brings a fresh approach and sense of humor to the matters at hand. For instance, at one point he engagingly leads readers through the details of interpreting a Smith Chart; at another, he leaves them admiring his conversion of a string of Christmas tree lights into a dipole antenna — with a visible standing wave.

The book then progresses beyond fundamentals into how the structure of the ionosphere directly affects radio propagation. The author spent decades exciting and probing the ionosphere and knows its behavior firsthand. Even the long-time Amateur Radio operator will benefit from his detailed descriptions of the ionosphere's many characteristics that affect radio transmission and reception.

Armed with the knowledge and skills imparted by this book, the Amateur Radio community may well contribute to the further understanding of atmosphere and ionosphere disturbances. Mr. Nichols puts forward many interesting ideas that use the very distributed nature of Amateur Radio operators and their equipment as a resource in the continued exploration of intriguing ionization events. I look forward to seeing how this resource, left underutilized by radio science, emerges and perhaps one day contributes to a first-rank scientific discovery.

David Saltzberg
Professor of Physics and Astronomy
University of California, Los Angeles
April 22, 2012

Preface — Unfinished Business

Part 97.1 of the Federal Communications Commission's rules and regulations defines who we are and what we do as radio amateurs. There are five elements listed in our basis and purpose section under Part 97.1. Right after element (a), which addresses our value as emergency communicators, we find,

> *(b): Continuation and extension of the amateur's proven ability to contribute to the advancement of the radio art.*

Somewhere in the recesses of most hams' memories this important passage lurks, but few hams have any idea what to actually do about it. We have been led to believe that one needs an advanced degree or a career in particle physics to really do anything technologically innovative, or that the radio art has been advanced as far as it can be advanced. This book, *Radio Science for the Radio Amateur*, will neatly demolish these misconceptions, along with a few others.

Although we know a great deal about radio in the 21st century, there is a lot we still don't know. Most of the unexplored frontiers of radio have more to do with the science of radio than with the technology of radio. There is an important distinction between science and technology, which is blurred in much literature, and frankly in most public policy. Science and technology are different but related disciplines. Radio amateurs tend to become infatuated with the technology of radio, while neglecting the far more interesting — to my mind — science of radio. As we move into the body of this volume, we will discuss more thoroughly the differences between science and technology: It is far too important and subtle to handle properly in this short preface.

Radio Amateurs Advancing Radio Science

One of the most dynamic fields in all of science is the search for new, exotic subatomic particles. Traditionally, this has been the domain of "Big Science" particle accelerators and atom smashers. And, of course, these large, expensive instruments will always play a huge role in fundamental science. But scientists also realize that many of these particles already exist outside the lab, "in the wild," as it were; we just aren't in the right place at the right time to see them. Back in 2000, I was privileged to participate with a community of scientists who convened at the

Radio Detection of High Energy Particles first international workshop. The RADHEP participants proposed many fascinating experiments using widely scattered arrays of simple radio receivers of various sorts to detect these particles naturally. Although the RADHEP proposals were directed at the attending scientific community, none of these receivers were beyond the skill level or resources of the average radio amateur. Modern data acquisition methods, as well as networking technologies we all now take for granted, easily allow hams to take part in "Big Science" projects — the sort of projects that result in Nobel Prizes, by the way.

Regarding the unfinished business of scientific inquiry and data acquisition, the participants at RADHEP suggested some of the most promising avenues of study, in one of the most widely open fields, for radio amateurs. One thing radio amateurs have a reputation for is making reality out of suggestions.

One Suggestion: Big Science on a Ramen Budget

I have an impoverished Amateur Radio friend who quips, "I have to eat Bottom Ramen, because I can't afford Top Ramen." Even during the best of times, some of the world's premier scientists have taken what amounts to a vow of poverty. Basic science is a hard sell these days, and even when there's money to spare, it's hard enough to explain to a skeptical public why we need to send probes to Mars. During bad economic times, it seems that interest in basic science evaporates, and yet, this is where Amateur Radio can shine.

For many years, I worked in ionospheric research, both at HIPAS (HIgh Power Auroral Stimulation) Observatory, now decommissioned, and at HAARP (High Frequency Active Auroral Research Program), both in interior Alaska. These facilities are definitely in the category of "Big Science." The construction of HAARP was completed just a few years ago with a $200 million budget. There are only a handful of full-scale ionospheric facilities on the planet, but for the most part, most new radio science in recent years has come from them. One of the things I learned while working at HIPAS and HAARP, however, was that a lot of the science done at these facilities could be done by radio amateurs — if you have enough of them. Even monster installations like HAARP or the EISCAT (European Incoherent SCATter) facility in Tromsø, Norway, can only be in one place at once. Hams are everywhere, and a lot of ionospheric research can only be done with widely scattered sensors, which hams are uniquely equipped to provide. In *Radio Science for the Radio Amateur*, I suggest that much of this research can be performed by the Amateur Radio community — all on a Bottom Ramen budget — and that we can contribute significantly to completing some long-unfinished business regarding understanding radio propagation.

But Where Do I Start?

I'm glad you asked. *Radio Science for the Radio Amateur* starts out with the basic concepts and disciplines of the scientific method, so we can all start out on the same page. Whether you've never set foot in a lab in your life, or you've been experimenting for decades, *Radio Science for the Radio Amateur* will lead you into some of the most fascinating aspects of Amateur Radio. I know you'll enjoy the journey.

Eric P. Nichols, KL7AJ
November 2012

Acknowledgments

First and foremost, I dedicate this work to God Almighty, Who made all things possible, as well as interesting. I am humbled and grateful that He has deigned to share a few of His secrets with us mortals.

I am indebted to my parents who gave me an engineer's mind, an artist's eye, and a fully functional set of curiosity glands, as well as a stable environment in which to pursue them…an increasingly rare gift in this age.

I wish to thank Dr. Alfred Y. Wong for his vision in creating HIPAS Observatory, which for 20 years nurtured countless numbers of new scientists, and a few old engineers, in the most unlikely of places. While, in some ways, HIPAS seems like ancient history, its contribution to the intellectual environment of Alaska and elsewhere will never be forgotten.

I need to express my thanks to a number of current and former UCLA Plasma lab, UCLA Physics and Astronomy lab, and HIPAS staff and "hired guns": Ralph Wuerker, who taught me everything I know about optics and lasers; Glen Rosenthal, who taught me how to program in C as well as how to set up a plasma chamber; Peter Cheung, who gave me some real interesting engineering problems to solve; Jackie Pau, who put up with me; Helio Zwi, who taught me the fine art of data acquisition; Mike McCarrick, who taught me a whole lot about ionospheric science, and David Saltzberg, who offered some great advice on this book. I also wish to posthumously thank David Sentman of UAF for his valuable advice and guidance on a wide variety of scientific matters.

Last but not least, I am grateful for the fine staff members of ARRL, who for the past 30 years have turned my chicken scratches into works of literature. In this vein, Becky Schoenfeld, W1BXY, and Edith Lennon, N2ZRW, have conquered the most difficult task of putting together this particular Magnum Opus.

Introduction

The Radio Amateur and Scientific Investigation

As mentioned in the Preface, one of the primary reasons Amateur Radio exists, according to FCC part 97.1, is to "...advance the state of the radio art." Many hams would like to take part in this enterprising aspect of the hobby, but don't know how to get started.

The radio amateur is in a position to perform meaningful scientific investigation in a way that is simply not available to many others. As mentioned earlier, you don't need a doctorate in physics to conduct meaningful radio science, but you do need to know some basic scientific techniques, common to every field of science. As a radio amateur, you already have a head start.

This is an exciting time to be involved in science and technology, perhaps the most exciting in history. It has been said that there are two kinds of people in the world: those who make things happen and those who sit around and wonder what happened. By contributing your radio skills to pioneering research, you can be a big part of making things happen.

Most discoveries occur in a laboratory of some kind. A lab may consist of anything, from a few feet of counter space adjacent to a sink, to a behemoth like the two-mile-long Stanford Linear Accelerator (SLAC) facility in Menlo Park, California. Common lab sciences include chemistry, physics, optics, electronics, biology, physiology, botany, geology, and computer science, but there is really no limit to what can be studied in the setting.

There are laboratories without walls. For many years, I was privileged to have worked in ionospheric research, which involved the study of the upper reaches of the atmosphere using high-powered radio signals. There are even certain types of experiments being proposed (especially in the field of particle physics) that use the entire Earth as a laboratory.

Despite the huge variety of labs, there are certain common procedures, or methodologies used in all of them. Good scientific method is required in every type. If you learn and adhere to good scientific method, you can position yourself on the cutting edge of radio science and research, either individually or as part of a bona fide research team. Many great theoreticians are not great experimenters and must rely heavily on the insight and practical "chops" of others, including radio amateurs, to make their ideas reality. Insight comes from experience, and experience comes through following established methods.

If you thoroughly absorb this book, you will be able to move effortlessly between the chemistry lab and the physics lab, or the electronics lab and the computer lab, and so on. You may even want to start a lab of your own — even if it's just a corner of your basement shack — to study something that no one else is studying yet. In any case, this book is for you.

As a radio amateur, you are probably champing at the bit to get into some actual radio science, and you may be wondering why I'm talking about all this chemistry and physics stuff. As you'll discover, the universe is astonishingly consistent, and good scientific method is good scientific method regardless of the specific discipline. I, therefore, want to present a coherent, comprehensive look at science before delving into radio specifically. To that end, the chapters of this book are organized to first offer a general introduction to the art and methods of scientific investigation, followed by a focus on radio science and how the radio amateur might participate in some important work in this intriguing field, and concluding with some specific suggestions to get you started in conducting solid science.

Please avoid the temptation to skim over the early chapters to get to the juicy, and perhaps more familiar, radio sections; you might miss some exciting and curious scenery on your way to the radio lab.

On Mindset, Measurement, and Making Good Science

The most important tool for the successful scientific researcher or investigator is a passionate desire for objective truth, regardless of the branch of science being pursued. As a scientist, amateur or otherwise, your job is to remove human opinion and bias from any scientific observations. Our senses are extremely unreliable; we are easily deceived by what we see, hear, smell, taste, and touch. Although good scientists use all their senses, they don't rely on them.

Unfortunately, much of Amateur Radio has been plagued by a lot of pseudo-science and even worse, some entirely unscientific folklore and opinion. I call this the Quack Factor. To avoid the Quack Factor and be a credible and reliable radio investigator, everything you observe must be confirmed through valid, verifiable, and repeatable measurements. You will take measurements by a variety of instruments, and anyone else using the same instruments will arrive at the same results. This is what good science is all about. Anything you discover in your ham shack, your workshop, or in your radio science lab, another scientist using the same methods should be able to discover in his or her shack, workshop, or lab. If not, it is bad science.

THE GIFT OF SUSPICION

One of my mentors in lab experimentation was proud to declare that he had a finely tuned gift of suspicion. The longer I've been in science, the more I realize how valuable — and rare — the gift of suspicion is. Unfortunately, in recent years within the Amateur Radio community, the gift of suspicion is not as finely tuned as it should be. We tend to see what we want to see, and sometimes our perception diverges from reality to a large extent. That discrepancy can have serious consequences.

For example, I've known a lot of aircraft pilots over the years and am currently working toward my own license. One thing you have to learn immediately for safe flying, especially if you may face bad weather as we do so often in Alaska, is the importance of doubt. A good instrument-rated pilot is trained never to trust his senses, no matter how convincing they may be. In bad weather, relying on your senses will get you killed flying in Alaska, or anywhere else for that matter. It takes a great deal of discipline to trust your instruments and not your subjective perceptions, your feelings, that is. Although a wrong conclusion in scientific research won't get you killed as easily as it will piloting a small plane, it can have longer-lasting effects. Stupid ideas tend to acquire long half-lives.

Good scientific method can only be objective. Physical reality itself is not subject to prevailing opinion or the particular interests of any person or group. In science, reality is straightforward: If you can't measure it, it doesn't exist. Period.

Does this seem restrictive and confining? Not at all; it is instead the foundation upon which we build our body of understanding. As we will learn, the first step toward new discovery is the ability to say, "I was wrong." Indeed, when a valid instrument indicates that you (and possibly everyone before you) were wrong, you may find yourself in the exhilarating position of gazing at an actual scientific discovery.

DISCERNING DISCOVERY FROM DUCK

In the laboratory, workshop, or other investigative setting, there is one instance where the exception may prove the rule and your senses can be fairly reliable, or at least trainable: recognizing the Quack Factor and nipping quack science in the bud. In this case, your experience may be the best instrument there is; in all other cases, we rely on precision instruments and established standards.

Some Terms of the Trade

Scientific instruments allow you to accurately and consistently measure any number of discrete values, such as the following:

- Acceleration
- Brightness
- Color
- Current
- Distance
- Force
- Hardness
- Loudness
- Mass
- Polarization
- Speed
- Temperature
- Time
- Voltage

This list represents just the smallest of samples. There are myriad values that can be measured, and as science progresses, new ones describing phenomena we're not yet aware of will be added. To be scientifically meaningful, these values must have some unit of measurement associated with them, with each unit being an accepted standard of some physical or theoretical mechanism. Common units include second, mile, gram, volt, ampere, and decibel.

Although some values may have multiple units associated with them — for instance, distance can be measured in miles or meters — there must be an agreed-upon standard for each unit to allow for comparison.

In the United States, the organization responsible for establishing and maintaining these standards is the National Institute of Standards and Technology (NIST), formerly the National Bureau of Standards (NBS). There is also an international bureau of standards, commonly known by its French acronym, BIPM, for the Bureau International des Poids et Mesures. The BIPM still houses an extremely precise "meter stick," an original prototype that was used worldwide as the reference for the exact length of a meter. This standard was recently redefined using an even more accurate electromagnetic measurement. The internationally accepted meter is known as a primary standard. Both the BIPM and the NIST maintain countless primary standards for just about any physical value you can think of, and many you've never thought of.

Since every science investigator who needs an accurate meter stick (or weight, or any other means of measurement) can't visit the NIST or BIPM personally, secondary standards were established. A secondary standard is a measuring instrument that can be accurately compared with, or traceable to, a primary standard. A well-equipped laboratory or workshop will have various secondary standards that are traceable to a national or international primary standard. Since standards and traceability are pivotal to good science, I encourage you to learn more about them. One recommended starting point is the "Supplemental Materials" section you'll find linked on the NIST web page located at **www.nist.gov/traceability/**.

Again, we aren't quite as concerned with the actual units that we use in research as we are with the agreement of those units. For valid comparison, my meter must be the same as your meter, my pound must be the same as your pound, and so on.

Some Tools of the Trade

Let's pull things together a bit by looking at the aforementioned values and connecting them with common instruments used to measure them and some common units used to describe those measurements.

Acceleration
Instruments: accelerometer, scale, balance, rate-of-change Doppler radar
Units: feet/second squared, meters/second squared

Brightness
Instruments: photocell, photoresistor, phototransistor, CCD (charge-coupled device), digital camera, photomultiplier
Units: lumen, watts/square centimeter, magnitude

Color
Instruments: color filter, diffraction grating, prism, spectrograph, spectroscope, pyrometer
Units: angstrom, nanometer

Current
Instruments: thermocouples, d'Arsonval meter, current probe, incandescent lamp
Units: ampere, milliampere, microampere

Distance
Instruments: ruler, meter stick, tape measure, odometer, interferometer, radar, lidar (light detection and ranging)
Units: centimeter, inch, yard, meter, mile, kilometer, micron, angstrom

Force
Instruments: scale, balance, load cell
Units: pound, Newton, dyne

Hardness
Instruments: Vickers test apparatus, Rockwell test apparatus, Knoop test apparatus
Units: HV (Vickers hardness), DPH (diamond pyramid hardness)

Loudness
Instruments: microphone, oscilloscope, sound level meter
Units: decibel (acoustic), phon, sone

Mass
Instruments: spring scale, balance, accelerometer, Mettler balance, gravitometer
Units: kilogram, slug, grain, metric ton, stone

Polarization
Instruments: polarized filter, polarimeter, dipole antenna
Units: degree, radian, pure number

Speed
Instruments: stopwatch and ruler, speedometer, interferometer, Doppler radar
Units: inches per second, miles per hour, astronomical units per light-year

Temperature
Instruments: thermometer, thermocouple, pyrometer, spectrograph, barretter, thermistor, bolometer, calorimeter
Units: degrees Fahrenheit, Kelvins, electron volt

Time
Instruments: sundial, chronometer, mechanical clock, atomic clock, electronic oscillator
Units: second, hour, month, year

Voltage
Instruments: d'Arsonval meter, oscilloscope, electroscope, digital voltmeter
Units: volt, statvolt

Again, this is just a small sampling. New tests and testing methods are being devised every day as we learn more about the universe. But don't be intimidated: there's only a finite number you'll encounter in your research.

GOOD SCIENCE BEGINS AT HOME

Step one is to start thinking like a scientist. All it takes is a little discipline and some basic training. This book begins with a discussion of some simple but effective methods for separating scientific fact from folklore, fantasy, and feelings. We will learn the rules of the road for scientific investigation in any specific discipline, and then narrow our focus on radio science, which involves all the physical phenomena in our world that can be explored by radio methods available to every radio amateur. In reality, *all* science is radio science to some degree, because radio waves permeate every aspect of the known universe — and, most likely, the unknown universe, as well.

Radio propagation is one of the most informative and simplest areas of radio science most hams can explore. It is already familiar territory for most of us, yet full of endless surprises. Using that as our cornerstone, we will then go much deeper than the surface effects of radio propagation and examine some of the plasma physics involved, much of which is still unexplored territory, taking a close look at some strange phenomena many of us have observed but maybe not understood. We will even explore some of the convincing evidence that some commonly observed radio disturbances may be caused by Earthly encounters with exotic particles.

Despite the surprises we encounter through radio science, the field demonstrates just how consistent the universe is. Indeed, we can use the same mathematics to describe everything from electrical resonance to the orbit of the planets around the Sun. A foundation of good science is the conviction that things can be known, even if they are a mystery at the time. Objects and events can be measured, results can be reproduced, and, eventually, phenomena can be understood. One important principle that will keep cropping up is *correlation*, the close tie between cause and effect. We will also often see the correlation between radio and other branches of science.

Simply stated, radio waves make things happen, and things that happen almost always create radio waves. For instance, chemical reactions cause electrical currents, and electrical currents cause radio waves. Also radio waves cause electrical currents, and electrical currents can create chemical reactions.

Now, with things chemical and electrical already on our minds, let's take the next step into doing science at home (or in the laboratory): learning to do so safely.

2 Safety First

Theoretically, any book on scientific investigation and experimentation should begin with this chapter. It takes second place here only because I think it's bad literary form to start out with a topic many people consider boring. That said, let's turn now to what must be your primary concern in the investigative setting: Safety.

Whether you conduct your scientific investigations in your basement shack or in a high-end research facility you must work within a safe environment. And, since safety first means being prepared, I'll take more of the high-end approach, offering information that applies to the professional environment as well to the ham shack and home lab. Much of it may not relate to you now, or possibly ever. On the other hand, one day you may find that you need to, say, etch a circuit board using some nasty chemicals and sharp objects. Perhaps you'll carry your pursuit of practical science even further.

NOT SO MAD SCIENCE

For many people, the word laboratory conjures visions of Frankenstein's lair, a dark, dangerous dungeon filled with vile potions, hissing chambers, and flaming blue electrical arcs. In rare cases, this classic image is actually correct (it was at the old UCLA Plasma Physics lab, where I worked intermittently for many years), but most modern laboratories take safety considerations seriously.

The hazards you may encounter in a laboratory vary widely. In a computer science lab the biggest danger you face may be carpal tunnel syndrome from constant typing, or clogged arteries from spending all day in your chair and ingesting only pizza and soda. In a destructive testing lab, on the other hand, you have to be alert to dangers such as flying objects, deafening sound levels, and possible electrocution. Of course, any modern lab takes these dangers into account and offers proper protection and training so that the risk to life and limb is minimized. Biological testing labs offer a set of hazards in a category of their own. Such labs have their own highly regimented safety programs, far beyond what we can detail in this book. It is unlikely, however, that you will be setting up a biological testing lab of your own. The bottom line here is that you need to carefully follow all the local and specific safety guidelines for whatever type of laboratory you may be called to work in.

Some basic precepts, however, are fairly universal. Let's look at these using the example of a standard chemical laboratory, which offers hazard levels about midway between those of a computer lab and a biological warfare lab.

TOP-DOWN SAFETY

We'll start near the top, with your eyes. I know two immensely capable blind lab technicians, but both were blind *before* they started working in laboratories. It would be tragic to become blind *because* of lab activities, but fortunately, this is totally preventable. There is a wide variety of comfortable and effective eye protection available, some of it even borderline stylish, in a nerdy sort of way (and looking nerdy is way better than being blind).

Don your eye protection *every* time you're in the lab, and don't take it off until you leave. Even if *you* aren't doing anything dangerous, other people in the lab may be doing dangerous or even *stupid* things. I remember a high school chemistry class where someone was making hydrogen by dropping a piece of zinc in a flask of hydrochloric acid. He was *supposed* to have inserted a vented stopper with a glass tube on the flask; instead, he just put a plain stopper on it. After simmering away in the back of the class for a bit, the flask exploded, sending hydrochloric acid and glass all over. Fortunately, everyone was facing away from the lab bench when this happened, so the damage was limited to a mess and several brief heart stoppages. There was no loss of life or limb, but that kind of inattention to safety protocol could have been tragic. Remember, *you* may be very smart and cautious, but your colleagues or fellow hams may not be! Enough said on that. Wear your goggles!

Depending on the materials you're working with, you may desire (or be required) to wear a mask of some sort. Not only may you be exposed to poisonous fumes, but breathing in dust and other particles, even seemingly harmless ones, may also present a danger. We all know about the hazards associated with asbestos, but particles from materials like talcum powder and chalk dust have also been found to pose health risks, especially over extended periods. It's a good idea to wear a mask if you're going to be working with any powders, whether they're known toxins or not.

Lab smocks are a great idea, too, and not just in a chemical lab, but in pretty much any kind of lab. They keep nasty chemicals from staining your clothes and offer a degree of protection for your arms. Don't rely too heavily on them, however; fabric smocks are porous and you might want long-sleeved neoprene gloves or other appropriate gauntlets for added safety.

Not long ago, gloves were strictly required only in certain situations, such as in semiconductor clean rooms and biological labs, but in recent years wearing gloves has become fairly universal in many kinds of labs. Modern latex gloves are thinner and a lot less annoying than they used to be, so there's really no reason not to get in the habit of wearing them.

From personal experience, I know how unpleasant it is to drip molten solder on a sandal-clad foot, and I'm assuming that pouring concentrated hydrochloric acid on a foot is orders of magnitude worse. So you never have to find out, wear fully enclosed shoes whenever you're working in the lab, whether it's required by company policy or not. You probably don't need steel toe boots in the average lab, but do wear real shoes.

VOLTS AND JOLTS

Nearly every lab involves the use of electricity, and some labs deal with nothing but electricity. In either case, you need to know some critical facts about electricity and its hazards. The main hazard, of course, is electric shock, or in its extreme form, electrocution, which is a fatal electric shock.

Contrary to what you may believe, it is not the voltage of the electricity that poses the danger: it's the current that determines its lethal potential. For example, on a cold, dry day, you can scuff your feet across a carpet and achieve a static charge of around 50,000 V. The "zap" you get when you touch a doorknob is annoying, but there's very little danger in it. On the other hand, you can easily electrocute yourself with a 12 V car battery if you were to connect it to probes attached to your opposite arms (and if you live on a high-sodium diet, your blood will be even more electrically conductive). Current flow is measured in amperes, and it takes only a fraction of an ampere to kill you. **Figure 2.1** is a chart taken from the Occupational Safety and Health Administration's (OSHA) website (**www.osha.gov/**) that outlines the relative effects of electrical current in terms of milliamperes, which are thousandths of an ampere. As you can see, it doesn't take much current to do severe damage.

Fortunately, every modern lab, and most modern ham radio equipment, makes it fairly difficult to electrocute yourself. There are, however,

Current Level (milliamperes)	Probable effect on human body
1 mA	Perception level. Slight tingling sensation. Still dangerous under certain conditions.
5 mA	Slight shock felt; not painful but disturbing. Average individual can let go. However, strong involuntary reactions to shocks in this range may lead to injuries.
6 -16 mA	Painful shock, begin to lose muscular control. Commonly referred to as the freezing current or "let-go" range.
17 - 99 mA	Extreme pain, respiratory arrest, severe muscular contractions. Individual cannot let go. Death is possible.
100 - 2000 mA	Ventricular fibrillation (uneven, uncoordinated pumping of the heart). Muscular contraction and nerve damge begins to occur. Death is likely.
> 2000 mA	Cardiac arrest, internal organ damage and severe burns. Death is probable.

Figure 2.1 — An Occupational and Health Safety Administration (OSHA) chart outlining how electrical current affects the human body. OSHA is an invaluable resource for safety information. [Courtesy of www.osha.gov/]

important precautions you should always take. For starters, whenever you're checking an operating circuit, keep one hand in a pocket, behind your back, or in your lap to avoid making a closed circuit through the body (**Figure 2.2**). Be sure that any electrical outlet near a water source has a ground fault circuit interrupter (GFCI)-type circuit breaker. Water and electricity do not mix, but a GFCI breaker will minimize the danger in case they get too close. Also, keep in mind than nearly every water-soluble chemical you can think of is even more conductive than water itself. Finally, be sure that items like hot plates, agitators, and other electrical equipment are well maintained and GFCI-protected. For more specific

Figure 2.2 — A safety-conscious electronics technician demonstrates safe use of instrument probes on unknown circuits. Notice the left hand is safely positioned far away from any source of electric current *or* electrical ground. [Eric Nichols, KL7AJ]

information, refer to the links you'll find at OSHA's web page at **www.osha.gov/SLTC/electrical/index.html**.

Should you find yourself in a modern dc electronic lab, be aware that it will have very high voltage power supplies, sometimes on the order of millions of volts. You may encounter a Van de Graaff generator, for instance, though its current capacity is extremely low, so the danger of electrocution is essentially nonexistent. Other common equipment includes insulation testing devices, "meggers" or "hi-pot" testers, which are capable of supplying on the order of 100,000 V, and significantly more current. These do present a degree of danger, and need to be handled with knowledge and care. You need not be terrified of high voltage dc, but you do need to know what you're doing.

Of the greatest danger, however, are power supplies in the 100 V to 1000 V range, which are easily capable of supplying lethal currents. You also need to be very careful of storage capacitors, which are capable of providing a lethal jolt. These can show up in unexpected places, such as photoflash units. They can retain lethal charges for weeks or months. Know how to safely discharge these devices.

Large batteries can be a hazard as well. Although you will rarely encounter batteries with enough voltage to present an electrocution hazard, the high currents available from large batteries can present a serious danger. A car battery is an effective arc welder, and any metal object you drop across its terminals can become red hot instantly. High current arcs from batteries and other power supplies can also create blinding flashes and spattering molten metal, with severe burn and fire potential.

ARCS AND SPARKS

A lesser-known danger posed by electrical apparatus in the lab is the possibility of explosion. Many electrical devices containing motors, relays or switches generate small electric arcs, just as a spark plug does, unless the appliance is rated explosion proof. Explosion-proof appliances place any spark-generating components in a sealed chamber.

There is also a slight danger in any lab that enough flammable fumes can build up to the point where such arcs could cause an explosion. Generally, this would occur only in a very confined area, but rare as such explosions may be, they can be very devastating. Any OSHA-compliant lab will have detectors and alarms warning of potentially explosive atmosphere.

Be aware that many labs have a host of Bunsen burners fueled by a common source of odorized (to make it detectable) natural gas. Bunsen burners don't have pilot lights, and it's fairly easy to leave one on overnight or longer. If you enter a lab and smell gas, no matter how faint, leave the room immediately, do not turn any lights on or off, and call the fire department and the lab safety officer. Although it takes a lot of gas to make a room explosive, it is always best to use an abundance of caution.

GOOD (AND BAD) HOUSEKEEPING

Since this book is about being at home in the home or lab science station, it's probably a good idea to cover some good housekeeping skills for that environment. Not only is a clean, orderly space more pleasant to work in, you'll also be more productive and much safer than you would be in a chaotic mess. While that UCLA Plasma Physics lab I mentioned previously was a fabulously productive beehive of science that yielded a continual stream of ideas, products, and a Nobel Prize or two, it could certainly have benefitted from some custodial care.

For both safety's and sanity's sake, keep floors clear, wipe up spilled hazardous chemicals (including coffee), wash your glassware, and put hardware away in the proper cabinets at the end of the experiment.

A dire warning is in order here, however. It is extremely bad form — and possibly dangerous — to put away someone else's hardware in a shared lab. I could tell you some horror stories about crucial experiments in progress being dismantled by a resident clean freak. In fact, it is best to *never touch* anyone else's experiment without explicit permission. Everyone in your lab will have a different operating style and personal space. Violate this at your peril!

You won't have to be too concerned about your own space being violated if your experimentation is done primarily in your home ham shack. There your biggest worry is probably your spouse cleaning up the wrong thing. To avoid that, I have a built-in "bat cave" in a corner of my garage where I perform most of my Amateur Radio and science activities. My family pretty much knows not to darken the door of the place, so my set-ups are fairly safe from meddling. Your mileage may vary.

KEEPING TOOLS SAFE, TOO

I want to close with some comments about equipment safety. While "Safety First" really means that you avoid personal injury, remember that there are a lot of expensive "toys" in any modern lab that you can damage. Use your common sense and always use the right tool for the job. For instance, don't use a five-pound soldering iron for miniature surface-mount components; this requires a delicate touch, and delicate tools.

There's a lot of science that needs to be done and we want you and your equipment to be around to do it.

The Basics

One of the rules of science is that you always study the *ideal* before you study the real. What does this mean?

We seldom encounter any substance or phenomenon in its pure state. For example, if you were a scientist attempting to learn everything possible about water, you'd need an absolutely pure sample of water to begin with. But even if you had a bottle of absolutely pure water, the second you opened its lid, you'd have smoke, smog and germs contaminating it. The smallest trace of certain compounds, such as salts, can drastically change the character of water from its purest form. Does this mean that it's pointless to attempt the study of pure water, knowing that you'll never, or at least rarely, encounter a sample of the stuff?

The answer is no, not at all. The ideal of absolutely pure water gives you the starting point — your *standard* — by which you compare every real-world example. Laboratory standards are the targets or models of perfection that we always strive to maintain in any measurements we make. Laboratory standards do indeed come very close to the ideal substance, and technology improves these laboratory standards every day.

ERROR AVOIDANCE

One of the main tasks of the serious scientist, amateur or otherwise, is to identify and minimize all possible sources of error. In fact, many common laboratory practices, in every field of science, are carefully designed to cancel out as many errors as humanly possible. This is best demonstrated with an example.

Let's say a research partner has asked you to measure out precisely 10 grams of sodium chloride for an experiment. You take a filter paper disk, a convenient vehicle for transporting small amounts of dry material, and place it on your balance. You spoon the sodium chloride onto the filter paper, carefully measuring out the 10 grams, and then transport it back to your associate. But she comes back a minute later saying she's short 1 gram of sodium chloride. How can this be? You carefully adjusted the scale and measured properly — but you forgot to weigh the filter paper. You should have zeroed out the scale with just the filter paper on it to ensure you only measured the sodium chloride. This zeroing process is known as error cancellation, and it takes into account a known value of error — in the case, the weight of the filter paper.

Or, if you're working in a quantitative analysis lab, you may be using a balance capable of measuring micrograms (millionths of a gram). When measuring such tiny amounts, you can actually spoil your measurement by a fingerprint left on a container (yes, you can measure the weight of a fingerprint, which is another good reason to wear gloves). The point here is that you have to pay attention to details, and the more advanced your experiment is, the more attention you have to pay to such details. Almost anything can cause errors in scientific measurements. How much it matters depends on what you're trying to accomplish, but it's best to get in the habit of striving for extreme accuracy. It will never hurt you.

Avoiding errors also requires you to pay special attention to the *conditions* of the experiment. For instance, one important measuring condition is temperature. Most substances have a density (mass per volume) that changes according to temperature, sometimes drastically. As a case in point, a pint of water weighs 1.04375 pounds — under the right conditions; that is, at *standard temperature and pressure* (STP), defined as 0° Celsius at standard atmospheric pressure. But water expands as it warms up, becoming less dense, so a pint at warm temperatures weighs less than it does at cold temperatures.

For accuracy, volumes of liquids must be measured out at known temperatures and pressures. The only reliable alternative is to measure out liquids by mass, which is not always as convenient as measuring by volume.

SOME STANDARDS, PLEASE!

As you develop your scientist mindset, you will realize there are numerous standards of measurements, such as the English system, metric system or imperial system. But don't despair; for nearly every scientific endeavor, the metric system is the standard. The metric system is especially handy for chemistry and electronics, because it is based on powers of 10, also referred to as orders of magnitude. There are, however, still instances where the English system is more convenient, particularly in the machining and manufacturing industries, and you should be comfortable working with various systems. The important thing to remember is that each of these systems is a *standard,* meaning there is a precise, well-known correspondence between them. As long as you know (and specify) which standard you are using, there will be no scientific error resulting from "translation."

Not only do you have to be careful about STP conditions while measuring liquids, but even distances can depend on temperature. A steel ruler (more accurately called a scale) may seem to be an extremely accurate distance measuring device, but in reality, it isn't. Steel expands as it heats up. It doesn't expand as much as copper and other metals, but it does expand. If you require extremely accurate distance (or dimension) measurements, you need to take this into account as well. Different materials have different *coefficients of expansion*, which is the percentage of change they undergo per degree of temperature change. Not only does a ruler or scale get longer when it heats up, it gets thicker and wider also; in other words, its total volume increases with temperature. This is defined as the *volumetric coefficient of expansion.* Again, your awareness of these considerations will help you avoid error.

To collect accurate data, whatever standard we use to measure must be far more accurate than the material or phenomenom in question and the conditions must be precisely controlled.

Here's an extremely useful rule of thumb: *Any measuring instrument you use must be at least 10 times as accurate as the item you are measuring with it.* It's not law, but it is a good guiding principle in any kind of scientific setting, whether it's your research ham shack or a physics lab. It's also accurate enough to keep you out of trouble in most cases, but not so stringent as to be unrealistic.

AS EASY AS FALLING OFF A LOGARITHM

This is probably a good place to introduce the topic of *logarithms*, which we will revisit many times throughout this text. The logarithmic function is one of those mathematical and physical concepts that just won't go away. Contrary to how it may have been taught in school, or how it's presented in even some of the best Amateur Radio literature, the logarithm wasn't invented to make life more difficult. In fact, its purpose is to make life simpler. We will find that in nearly any realm of scientific investigation we will encounter huge ranges of physical values we will need to work with. In radio, we will encounter power levels ranging from the minuscule to the immense. Expressing and calculating such vast ratios

Minimalist Receivers

To demonstrate the power of the decibel in my ham radio classes, I use a very practical example. I ask the students, "How much amplification (gain) do we require from a receiver in order to hear (without too much strain) a 'typical' Amateur Radio signal?" This is a great starting point for our discussion, too, because it helps us acquire a 'feel' for just what kind of power levels we're dealing with in Amateur Radio. Let's break the problem down, using some decibels where applicable, to show how this mathematical tool can simplify a lot of things. I find it useful to work from the back end (the loudspeaker) toward the front end (antenna) to answer this question.

The decibel was contrived by the Bell Telephone labs in the early days of the telephone in order to determine optimum power levels for telephones. Their researchers found that, with the typical telephone headset, a comfortable listening level for most customers occurred with an electrical signal power of −13 dBm. For nearly a century, the −13 dBm has been the benchmark for telephone circuit testing and such. Let's look at this number in detail.

First, what is that "m" at the end of dBm? Remember, the decibel is a power *ratio* — in itself, it tells us nothing about absolute power. We need a suffix or subscript to set a reference point. In this case, the reference is the mW, or 1/1000 W. Working through decibels backwards, we find that −13 dBm translates to 1/20 mW, or 1/20,000 W. This is for a clearly audible voice signal. So we have a goal of −13 dBm. Remember that number. The radio receiver in question has to increase the antenna input signal from whatever it is to about −13 dBm. The total gain necessary can be distributed through a combination of RF and audio amplification in the receiver.

Now, let's look at the incoming signal at the antenna input to a receiver. For most communications receivers, an "S9" signal (strong but not blistering) corresponds to 50 µV. We have to twiddle this figure a bit before we can do much with it, however. We need to know how much *power* a 50 µV signal represents. Most modern amateur communications receivers use a 50 Ω input impedance, or something reasonably close to that value. We can use the following formula to calculate the input power of a 50 µV signal:

$$P = E^2/R \qquad \text{(Eq 3.A)}$$

where
P = power in watts
E = voltage in volts
R = resistance in ohms

50 µV is 0.000050 V. If we square that, we get 0.0000000025. If we divide that by 50 we get 0.00000000005 or 5×10^{-11} W. That's 50 pW (picowatts)! Keep in mind, this is for a rather *strong* signal. If we're listening to signals down near the noise floor, they will be on the order of femtowatts (10^{-15}).

So, how many dBm is 50 pW, anyway? Well, one trick is to convert everything to its fundamental unit first. Although dBm is a common unit, it's still a *derived* unit; 0 dBm is actually −30 dBW, where dBW is referenced to a watt. How do we arrive at that? The answer to this demonstrates precisely why decibels are so convenient.

Using our basic definition of the decibel, let's convert from watts to milliwatts. If our "output" power is watts, and our input power is milliwatts, then we can express the power ratio as 0.001/1. The log of 0.001 is −3. It's a negative number because we have a loss in power. (In this case, the "loss" is merely changing to a larger unit). Multiply our −3 by 10, again in accordance with our original definition of the decibel, and we have −30 dBW. Hold that number.

Now, isn't −30 a much more convenient number than 1/1000? The first significant digit in any decibel reading is simply the number of decimal places you move the thing. Therefore, −3 means you move the decimal place three places to the left. Watch how this works with *very* small numbers.

How many dBW is 5×10^{-11} W? Well, let's just take the log of 0.00000000005 and multiply it by 10. That comes out to −103 dBW, plus some spare change. Remember that number, too.

Now, returning to our original target power of −13 dBm, we can convert that to dBW by simply adding our −30 dB to come up with −43 dBW. So our target output audio power is −43 dBW.

How much amplification to we need to get from −103 dBW to −43 dBW? We simply subtract the two figures, and it looks like 60 dB. Now, notice that we didn't say we needed 60 dB*W* of gain. Why not? Because when the input and output powers are in the same units, dB becomes a *dimensionless* value. We only apply the suffix to "anchor" it to an actual power value. When using dB as a comparison alone, we don't need a suffix. In fact, the statements "dBW of gain" or "dBm of gain" are wrong; 0 dBW is an actual power value — 1 W.

So, how much power gain is 60 dB? We can run the dB formula backwards and discover that 60 dB is a 1 million to one power ratio. This is the power gain you need to build an Amateur Radio receiver that will actually *work*. Does it sound like a daunting task?

Not at all. This can be done with a grand total of three off-the-shelf (or off-the-wall!) transistors, if you have only a modicum of design skill. The theoretical maximum power gain for a standard junction transistor is about 23 dB, but with moderate construction skill, any radio amateur can get 20 dB out of a transistor stage. Again, 60 dB is the total gain, distributed through the receiver. Traditionally, most of this gain was achieved in the RF stages, but in modern direct conversion receivers, nearly all of this gain may be achieved in the audio stages. We'll talk about direct conversion receivers in some detail later in this book.

Aside from the decibel's ability to make unwieldy numbers quite manageable, we find that in complex cascaded amplifiers, the total gain becomes a simple *addition* problem, rather than a multiplication problem. And from a larger, systemic, point of view, the decibel is even more useful. Most communications systems have a number of components, some with gain, such as amplifiers, and some with loss — such as free space propagation loss or transmission line loss. Decibels allow you to simply add all the gains and losses together and come out with a simple composite gain figure for any radio path or circuit.

of power can result in some hard-to-manage numbers — both for humans and for machines.

One of the more common applications of the logarithm in radio is the decibel. If we want to express power ratios in terms of decibels, we use the standard formula,

$$dB = 10 \log_{10}\left(\frac{P_2}{P_1}\right) \qquad \text{(Eq 3.1)}$$

where
P_2 = the output power of any device
P_1 = the input power to the same device

Using a common log (base 10) is standard practice in radio work, but we also could use the *neper*, a *natural logarithmic* decibel (the common logarithmic decibel is in keeping with our concept of orders of magnitude, however). It should be noted here that while base 10 is very convenient for a lot of science, it is not reflective of physical nature itself. The natural logarithm is more fundamentally linked to physical processes. For example, the familiar *resistor-capacitor* (RC) time constant uses the natural logarithm to find the voltage of a capacitor after a certain charge or discharge time.

So, contrary to our beloved decimal system, there is nothing especially revealing about powers of 10 from a purely physical standpoint, but they do make things convenient for us humans. (For more on logarithms, decibels, and what they mean for a radio receiver, see the "Minimalist Receivers" sidebar.)

IT'S ONLY NATURAL

If you look at nearly any phenomenon around you, you will find the natural logarithm at work. If you live in a house with other occupants, especially noisy ones, you probably hear them. If you don't want to hear them you go into your room and shut the door. You probably still hear them, but not quite as loudly. If you *really* don't want to hear them, you stuff a blanket in the crack under your door. You probably still hear them, but just a little more quietly. So you screw a couple of earplugs into your ear canals — and if you're really lucky, you might not hear them at all.

What you've demonstrated is the natural logarithm in all its glory. Nothing ever really ends, once it's been created — it just sort of fizzles out. Whether it's sound or light or radio waves, or radioactive decay, there are functions of intensity that follow a natural logarithmic curve of some sort, which in its simplest expression is just 1/X, where X is some distance from the starting point, either in distance or in time.

If it weren't for the natural logarithm, life would be very abrupt. Radio waves would travel a certain distance and then stop, sound waves would simply cease at some given distance from their source, cars could not accelerate or decelerate smoothly — indeed, no phenomenon would ever attenuate. If you dropped a basketball on a wooden floor, it would bounce forever, its echoes ricocheting off the gym walls through eternity.

So, the natural logarithm is your friend, even if the calculations necessary to work with it can be a bit difficult. It leads us into calculus, but we can turn most of the gnarlier tasks involved over to home computers. I do want to point out, however, that although we can avoid a lot of the mechanics of calculus with judicious use of computer power, we can acquire a lot of intuitive insights into the physical universe by investigating some calculus principles.

While we won't explicitly "do calculus" in this text, some calculus will be incorporated painlessly into various formulas and equations we present.

4

Data Acquisition — The Road from Concept to Confirmation

Up to this point, we have rambled through a fairly theoretical landscape; now let's follow the path of scientific method to travel from theory to real-world results. In this chapter, our starting point will be an overview of some fundamental concepts of electronics, which we'll illustrate with a simple, but enlightening, demonstration. From there, we'll introduce the crucial next step for the investigative radio scientist: *data acquisition* (DAQ). Of course, we'll need tools to help us along our way, but luckily most radio amateurs will already have them, or at least have easy access to them.

We're also fortunate in our task because electronics is a simpler subject than some other sciences in that it is extremely *linear*. What do I mean by this? Put simply, I mean that there is a very close connection between cause and effect in electronic apparatus, as compared to, say, meteorology or aeronautics, which are highly dependent on chance or unknown variables. Air, for instance, is extremely compressible, greatly adding to the complexity of aeronautical study; electricity, on the other hand, is totally non-compressible — it behaves very predictably most of the time, under a huge variety of conditions. For the most part, electronics is pretty much like plumbing: it responds directly and immediately to your commands unless something breaks.

To demonstrate how we can use electronics to measure any physical property, we need to explore a few fundamental electrical relationships. We will describe electricity in some very rudimentary, mechanical terms, precisely because this method works so well. In fact, if you understand plumbing, you already understand about 90% of all you need to know about electronics instrumentation for radio science. From there we'll dive right into an informative demonstration for you to conduct in your radio lab.

PUT IT IN WRITING

It should go without saying (but we won't make that assumption) that the scientific experimenter has to keep copious and accurate notes throughout each stage of his or her work. Starting with the "chicken scratches" you jot down during the initial inspiration phase, through the emergent thesis (your creatively evolving proposal) and accumulating measurable findings, to your analysis and conclusions you need to carefully document everything you do. Trust me, you'll wind up with a whole set of notebooks crammed full. In fact, you may want to keep two sets of lab notebooks: one that nobody else will ever see and one (neater) for public scrutiny.

It is one thing to perform an accurate, well-conceived experiment. It is quite another to meaningfully document your results. Since time

	A	B	C
1	DATE/TIME	5/4/2002 1100 ADT	
2	PAGE NUMBER	1	
3	TITLE	USING X AND O MODES FOR IMPROVED RADIOCOMMUNICATIONS	
4	HYPOTHESIS	X AND O MODE RADIO PROPAGATION IS MORE THAN A LABORATORY CURIOSITY. JUDICIOUS USE OFR CIRCULARLY POLARIZED ANTENNAS CAN GREATLY IMPROVE H.F. RADIO COMMUNICATIONS	
5	EXPERIMENTAL SETUP	SEE BLOCK DIAGRAM ON PAGES 2-4	
6	COLLECTED DATA (RAW)	SEE PAGES 5-6, TABLE 1	
7	ANALYZED DATA	SEE TEXT PAGES 7-14 AND ASSOCIATED TABLES	
8	CONCLUSION	THE PHYSICAL MECHANISMS OF X AND O RADIO PROPAGATION MODES HAVE BEEN UNDERSTOOD FOR OVER 70 YEARS, BUT HAVE NOIT BEEN IMPLEMENTED IN PRACTICAL RADIO COMMUNICATIONS. STATISTICALLY SIGNIFICANT IMPROVEMENT IN SIGNALTO NOISE RATIO BY MEANS OF CIRCULAR POLARIZATION OF H.F. COMMUNICATIONS HAS BEEN DEMONSTRATED IN CONTROLLED FIELD TESTS PERFORMED IN THIS EXPERIMENT.	
9	SUGGESTED FURTHER EXPERIMENTS	FURTHER EXPERIMENTATION OVER LONGER TIME FRAMES AND LONGER PATHS IS SUGGESTED, AS WELL AS THE USE OF HIGHER GAIN ANTENNAS THAN THE SIMPLE TURNSTILE ANTENNAS USED IN THIS EXPERIMENT.	

Figure 4.1 — The lab report is far more than a mere janitorial duty, akin to washing your glassware at the end of the day. In fact, this sometimes-onerous task can justify your existence as a lab technician, or even as a scientist. [Courtesy of Eric Nichols, KL7AJ]

immemorial, laboratory walls have been graced with reminders to the effect that "No job is done until the paperwork is completed." Your paperwork will be a work in (and of) progress, but visualizing a succinct lab report (**Figure 4.1**) may help your organization process.

THE PRESSURE'S ON

The first concept we'll look at in electricity is electrical *pressure*, more accurately known as *electromotive force* (EMF). Using our plumbing model, your home's water pipes have a constant pressure of about 30 pounds per square inch (psi) in them, with all the faucets turned off. This pressure is reduced once you open a faucet, but let's look at the "all faucets off" condition. This pressure in your water pipes has the *potential* of doing some work, but it's not doing anything at this time. This is the same condition you have with a battery with nothing connected to it. In the case of a car battery, the terminals present an EMF of about 12 V. A volt is the electrical equivalent of pounds per square inch. It is a potential, but it doesn't do anything until something else happens. That something else is *current flow.*

If you open a spigot, the 30 psi of water pressure will cause the water to flow at a certain *rate,* such as gallons per minute. If you have a hose connected to the spigot, you can use this power (water pressure times water flow) to perform a task such as washing mud off your car. The amount of useful work the water can do is the product of the pressure and the amount of water flow. If you turn off the spigot, the pressure will still be there in your house's pipes, but you have no water, thus no mud washed off. On the other hand, if you take a bucket full of water and pour it on your car, you have a lot of water flow, but no pressure — so the car remains muddy. You need both pressure and a quantity of water flow to get any useful car-washing done.

The electrical equivalent to water flow is *current.* The unit of electrical current is the *ampere.* In general, big fat wires can carry more current for a given amount of voltage, just as a fire hose can pump out a lot more water than a garden hose, for a given pressure.

Electrical power, which is the product of voltage (pressure) and current (flow), is measured in watts (1 A times 1 V is 1 W). Pretty simple.

OHM'S LAW

There is another crucial relationship to understand, and once you've got this down, you can do almost anything within electronics. This relationship is known as Ohm's Law, and it describes the connection between pressure and current with a given size wire (the hose in our car-washing example). Ohm's Law is simple:

$$E = I \times R \qquad \text{(Eq. 4.1)}$$

where
- E = voltage in volts
- I = current in amperes
- R = resistance in ohms

R can be thought of as the diameter of our hose. A high value of resistance is equivalent to a skinny hose, and a low value of resistance to a fat one. Ohm's Law tells us that for any given size of wire (hose), the current flow is going to be directly proportional to the voltage (pressure). If you double the volts, you double the amperes; if you keep the voltage constant but reduce the resistance by half, the current will double.

Ohm's Law has to be modified a bit to handle ac electronics, but for dc everything is pure and simple, as we shall see.

MEASURING UP TO DC

There has traditionally been a division of electronics into two main areas: dc electronics and ac electronics. Although the divisions are somewhat flexible, dc electronics generally involves voltages that remain fixed, or change very slowly. On the other hand, ac electronics involves electric currents and voltages that vary on a regular basis, generally, but not always in a sinusoidal (smooth repetitive wavelike) manner. It includes everything from 60 Hz commercial power transmission, up through radio frequencies and microwaves, and sometimes visible light and beyond. The math involved with ac electronics of any frequency is seldom more complicated than trigonometry, while dc electronics math is generally even more straightforward.

The first thing you will encounter in a dc electronics workbench or lab is a source of dc power, usually many of them. Batteries are a convenient source of relatively low dc voltage, but they have to be recharged and their voltages are not adjustable. We usually need to have a widely variable source of dc voltage, sometimes with different polarities. A modern bench-type power supply is a versatile instrument, capable of supplying a wide range of voltages, perhaps several simultaneously. Most of these also have current limiting and over-voltage protection to prevent you from frying your projects in case you miswire them (trust me, you will). There are also constant current power supplies, which can be quite handy for many experiments.

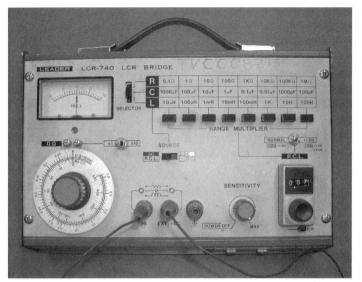

Figure 4.2 — A general-purpose component bridge is a valuable instrument in any electronics lab. These can be picked up surplus at very reasonable cost. [Eric Nichols, KL7AJ, photo]

Figure 4.3 — The quartz crystal is a central component in nearly every modern radio circuit and computer. You can even use your shack's analog bridge for testing these. [Eric Nichols, KL7AJ, photo]

Not only do you need to supply electrical power in an electronics lab, you also need to be able to measure it accurately. There are numerous devices available for measuring dc voltages and currents. The digital multimeter (DMM) is a versatile instrument capable of accurately measuring voltage and current (and often other electrical properties) over a fairly wide range of values. You can find an inexpensive, yet astonishingly accurate, DMM at your local hardware store for under $20. The main difference between an expensive, professional-quality device and a cheaper one is that you can run over the former with a pickup truck and it will still work.

As all-purpose as the DMM is, there will be times when you'll need more specialized instruments, such as when measuring extremely high or extremely low voltages or currents. Also, as you start to interface experiments to computers, you will need some analog measuring tools. One of the oldest and most useful of these is the bridge, which comes in countless varieties and configurations (**Figure 4.2**) and can be used to take accurate measurements of components as diverse as transformers and crystals (**Figure 4.3**).

A SALTWATER DEMONSTRATION

As we mentioned in Chapter 1, there is a very close association between electrical and nearly all other physical processes, and we can learn a lot about both through simple manipulation and observation. Chemical reactions are primarily the result of electron activity; in fact, every chemical reaction not only creates an electrical current, but also responds to an electrical current.

The most obvious example of this interaction is the rechargeable battery, such as that heavy lead-acid one under your car's hood. The battery, of course, supplies electrical current as a result of chemical activity, but we can alter the chemical makeup of the battery by forcing electrical current into it, as during the recharging process. An even simpler process than the lead-acid battery reaction is water *electrolysis*. By passing an electric current through an electrolyte, a chemical compound that dissociates in solution into ions — such as saltwater — we can observe chemical changes. For instance, if we force electrical current through water, we break the water down into its two elemental components: hydrogen and oxygen.

We can even measure the amount of salt in a sample of saltwater by measuring the electrical resistance (or, more accurately, the resistivity) of the sample. Pure distilled water has nearly infinite electrical resistivity, while brine is a very good electrical conductor. In fact, electrical resistivity is a very sensitive and accurate indicator of the salinity of water. It would be extremely difficult to obtain the same level of accuracy by measuring the mass of salt in the solution.

Let's put together a simple demonstration that illustrates basic electronics principles. In conducting it, you will manipulate Ohm's Law in a few different ways to better understand the concepts behind it. We'll also use it to explore the all-important task of *data acquisition*, or DAQ. All you need is readily available in any proper electronics workshop, and a lot of it right in your kitchen:

Demo 4.1 — Materials and Procedure

To begin this demo, you'll need:
- Adjustable 0–30 V dc power supply (a "wall wart" with a multiple output voltage switch serves nicely)
- Several insulated alligator clip leads
- Collection of ¼ W resistors: one 1000 Ω, one 10,000 Ω, and one 100,000 Ω (exact values are not crucial; we will actually measure them first)

- Digital multimeter
- 1 quart of distilled water
- Graduated beaker
- Box of sodium chloride (un-iodized salt)
- Two 1 × 3 inch strips of aluminum foil
- Hot plate or coffee re-warming plate
- Lab thermometer or candy thermometer
- Measuring spoons

Fill the beaker to the top graduation with distilled water. Take one of the aluminum strips and place it into the beaker right against the glass so that precisely 2 inches of the foil is below the surface of the water. Take the top edge of the foil and fold it over the lip of the beaker. Clamp the foil to the lip of the glass where it folds over with an alligator clip lead. Take the second aluminum foil strip and place it into the beaker exactly opposite the first one, this one also protruding 2 inches into the water, identical to the first one. Attach one end of another alligator clip lead to this strip in the same manner (using different color alligator leads helps you keep track of things).

Make sure your power supply is OFF (you may turn it back on after everything is connected up). Take the free end of your first alligator clip lead and connect it to the *positive* (red) terminal of the power supply. Take the free end of the second alligator clip lead and attach it to one lead of the 100,000 Ω resistor. Take a third alligator clip lead and attach it between the second lead of the resistor and the *negative* (black) lead of the power supply. You now have a simple series circuit.

Electric current must pass through both the water and the 100,000 Ω resistor to form a complete circuit. In an electric circuit, electrons always pass from the negative terminal of a power source toward the positive terminal. In this setup, electrons are going to flow from the negative terminal of the power supply, up through the resistor, through the water, and then into the positive terminal of the power supply. We will also have some ions, dissociated water molecules, passing through the water itself in the *opposite* direction (we can ignore the ion flow for now).

THEORY BREAK

Let's pause for a moment to add a few crucial pieces of information we need to know about electrical circuits:

1. In a series circuit, such as our present setup, the current (amperes) is the same in any part of the circuit. Exactly the same amount of current will flow through the water, or through the resistor, or through the power

supply. In fact, this is the definition of a series circuit; there is only one path or loop through which the electrons can flow.

2. Resistances in series always add. In the case of our experiment, we have two resistances: the actual physical resistor (100,000 Ω) and the resistance of the water, which is unknown (X). We'll call the total resistance 100,000 + X.

3. In a series circuit, it's the *total* resistance that must be inserted into Ohm's Law to make it work.

With this in mind, let's see what we can learn about this circuit. First we'll look at what we *don't* know, and then see what we can do to remedy the situation. Again, X is the resistance of the water and is what we don't know. Let's see how Ohm's Law fits our situation.

Looking at the "raw" equation (E = I x R), we first encounter E, which is voltage. Do we know what it is? The answer is yes, because we're going to intentionally set it to 10 V once we turn on the power supply. This makes E an *independent variable*.

How about I, the circuit current? No, we don't know that yet, but once we supply some voltage, we can measure it. How?

Let's rewrite Ohm's Law to express I in terms of what we *do* know, or rather in terms of what we *can* know.

I = E / R

It's the same equation showing the same relationship, but with different independent variables. Continuing through the equation, do we know the total resistance? No. Do we know any part of the resistance? Yes. The known part of the resistance is 100,000 Ω. Once we turn on the power supply, we can measure the voltage across R1 and thus determine the circuit current. But how can we do this, if we don't know the total resistance?

Remember what we said about series circuits, that the current is the same everywhere? If we can figure out the current through the resistor, we know that the current through the water has to be the same; in fact, we don't even need to know the power supply voltage to figure this out (we will need this later, however).

Let's turn on the power supply now and see what happens. Just to be precise, let's take our DMM and measure the voltage right at the power supply terminals. Adjust the output voltage to read exactly 10 V. Next we'll measure the voltage across the 100,000 Ω resistor, our *known* component. Let's say we read 0.1 V across the resistor. Can we figure out the circuit current? Sure. Using I = E / R, we get I = 0.1 / 100,000, which equals 0.000001 A, or 1 μA. That's not very much, but it's what one

Practical KVL

Sometimes confusion about what we mean by "voltage drop" arises when doing Kirchhoff's voltage law measurements. Concisely defined, *voltage drop* is the *difference of potential across any single component.* However, the term itself doesn't adequately describe the *polarity* of the difference of potential, which is something you need to know to make KVL "work." Here's a neat little trick that works as a "thought experiment" as well as with real measurements. When measuring voltage drops around a single loop containing several resistances, and perhaps even several voltage sources, remember this rule: *Red Leads Black.*

What this means is, always keep the red lead of your voltmeter *in front of* the black lead of your voltmeter, as you follow the various voltage drops around the loop. This will *automatically* reverse the polarity of your meter probes at the appropriate times, so that the voltage drops equal zero. This is a somewhat more reliable approach than "tweaking" KVL to say that "the sum of the voltage drops in a series circuit equals the supply voltage." Although this is true and standard practice for most electronics technicians, it doesn't work so neatly when you have, perhaps, multiple voltage sources, perhaps of different polarities. Let's say you have a series circuit consisting of a 9 V, a 10 Ω resistor, a *reverse polarity* 6 V battery, and a 20 Ω resistor. What is your supply voltage in this case? If you always follow the *Red Leads Black* rule, you'll always have the correct answer: the sums will always equal Zero.

You may wonder just *why* anyone would create such a circuit in the first place —one containing two opposite polarity power supplies. You generally wouldn't do this intentionally, but many complex circuits have voltage sources that are not obvious. For example, in a transistor or vacuum tube circuit you may have a negative *bias* voltage as well as a positive collector voltage (or *anode* voltage for a tube) on the same device.

Well Grounded in KVL

As a practical note, electronics technicians generally measure voltages *with respect to ground*, rather than voltage drops across individual components. But where is ground? As far as KVL is concerned, there's nothing sacred about "ground potential." It can be chosen arbitrarily. In our example above with the batteries and resistors, we can assign any of the nodes (junctions between any two components) as ground. KVL doesn't know or care where ground is. But, a lot of instruments *do* care, such as a typical oscilloscope or RF signal generator. In addition, circuit modeling programs such as *SPICE* (which we discuss in a later chapter) care deeply about where ground is.

It behooves you to become familiar with both ground-referenced and "floating" voltage, or "difference of potential" measurement. In fact, most data acquisition (DAQ) devices allow you to measure either KVL-friendly voltage drops using the *differential mode* connections or ground referenced connections. You will most likely end up using both at various times.

might expect for pure distilled water, which is a pretty lousy conductor.

Now that we know the total current, we can work backwards and figure out the resistance of the water. Ohm's Law, performing one more somersault, gives us: R = E / I.

Again, it's the same equation, just expressed differently. In this case R is the *total* resistance so we need the *total* voltage. Do we know that? Yes, we set it to 10 V. Do we know the total current? Yes, it's 0.000001 A. So 10 / 0.000001 = 10,000,000 Ω (or 10 MΩ). But that's the *total* resistance; now we need to subtract our other resistance to come up with just the water resistance. So we have X (water resistance) = 10,000,000 − 100,000 Ω, which is 9,900,000 Ω (or 9.9 MΩ). Nothing to it! As convoluted as this process may seem, you'll soon be able to do this sort of thing in your head. It just takes a bit of practice.

We've just seen Ohm's Law in all three "flavors" in this exercise, all of which repeatedly show up in electronics problems of any sort. We can also double check our work with a few other methods, as we'll soon see.

Another characteristic of a series circuit is that all the voltage drops across each component add up to the supply voltage, which is simply the difference in voltage between the two terminals of any component. This is a corollary of Kirchoff's Voltage Law (KVL), one of two fundamental laws in electrical engineering, the other being Kirchhoff's Current Law (KCL). KVL tells us that the sum of all voltage drops in a closed loop equals zero. How do we verify this? Well, if we measure the voltage across the resistor and add it to the voltage across the water, it should equal our supply voltage. So, if you take the DMM and measure between the two aluminum strips, you should get 10 V − 0.1 V, or 9.9 V. Pretty slick.

Despite all we've learned so far, we still haven't gotten to the main point of the experiment. We've merely set up some valid instrumentation and have a certain degree of confidence that what we measure next is somewhat believable.

EXPANDING OUR DEMO

Now that we're a little wiser, let's find out what salt does to the electrical resistance of water. We'll add a teaspoon of salt to the beaker, stir it up thoroughly, and measure resistance. Actually, we'll *measure* the current, but *calculate* the resistance. Again, if we measure the voltage across the 100,000 Ω resistor, we can know the total current. Once we know that, we can know the total resistance. Then we can subtract our 100,000 Ω and come up with the water resistance.

We can be tricky about this, too. Connect the DMM permanently across the 100,000 Ω resistor using a couple more alligator clip leads. Keep adding salt until the meter reads 5 V. What does this tell us? It tells us that the water resistance is the same as the resistor resistance, or 100,000 Ω. Half the voltage is dropped across the resistor, and half is dropped across the water. (I trust you kept careful track of exactly how much salt you added to achieve this condition, and I suspect that it took a lot less salt than you would have guessed.)

Let's pause at our 100,000 Ω mark for a while. Put the beaker on a hot plate and bring the temperature up to 180°F or so. What happens to the water resistance? Is it what you expected? Now, we can do something really radical. Carefully measure the voltage across the water and record that reading. Next, crank up the power supply voltage to exactly 20 V. Don't just rely on the power supply's meter, if it has one; re-measure to verify the voltage at the power supply terminals. Go back and measure the voltage across the water. According to Ohm's Law, you should see twice the reading across the water as you did before. Yes? Good!

Note, though, that there are actually a few cases where this might *not* work. Many electronic devices are *non-linear*, meaning the resistance changes depending on how much voltage is applied. In fact, most semi-conductors are non-linear in this regard. It doesn't mean that Ohm's Law doesn't work for non-linear devices, just that we have two independent variables. If we know what those variables are, we can still use Ohm's Law. Saltwater is quite *linear*, and the resistance is constant for any applied voltage, assuming the temperature and salt concentration is held steady. But the relationship between resistance and temperature is *not* linear. If you took several readings of resistance at different temperatures and plotted them on a graph, you would see this non-linear relationship.

Back in the old days, we used a lot of graph paper; today the trend is to plot everything on computers. Plotting points on a graph is extremely useful in every science as it allows us to see patterns and to make sense of what can seem to be random events. It's one aspect of the critical process of scientific discovery and leads us very smoothly into our next topic: DAQ.

APPLYING DAQ TO OUR DEMO

Although data acquisition is commonly considered part of computer science, it's actually as old as writing down at lot of numbers and scratching your head over them. To be more precise, the "scratching your head" part falls into the category of data analysis and is the "making sense" part of the whole DAQ process. Computers help you do both.

Since we have such a nice experiment going already, let's take a look at how we would do some data acquisition and analysis the old-fashioned way, and then show how we can modernize and automate the entire process with a computer.

Demo 4.2 — Expanded Materials and Procedures

Let's take our investigation a little further by adding:
• Quadrille-ruled graph paper
• Shack or lab computer capable of running spreadsheet program

First we'll turn off our hot plate and let our sample of saltwater return to room temperature. This time, we're going to take many resistance measurements as the water is reheated from room temperature up to the maximum temperature the hot plate can give us. Since we've already established the relationship between resistance and voltage drop, we don't have to calculate the resistance after each voltage measurement; we can do that later, or better yet, let a computer do that for us.

Reconnect the DMM across the fixed resistor. Leave the supply voltage at 20 V, as that will give us a little better sensitivity. Once you're confident the saltwater is at room temperature, take an initial voltage reading with the DMM. Now turn on the hot plate. For each 10° of temperature increase, write down the voltage reading. You will notice (if you didn't glean this from the previous demonstration) that the circuit current goes up with temperature, which means the saltwater has less resistance or better *conductivity* with higher temperature. This, by the way, is the opposite of copper wire, as well as most metals, which has a higher resistance with higher temperature. This is known as a *positive temperature coefficient*. Saltwater has a *negative temperature coefficient*, as do semiconductors.

After the hot plate reaches its maximum temperature and you have all your voltages and temperatures scribbled down, we can begin to *correlate* them. The simplest, old-school correlating tools are a pencil and graph paper. Plain old quadrille-ruled paper (⅛ inch squares are fine) will do this job nicely. On the horizontal axis, indicate each temperature; you can use one square for every 10°, or two or three squares for every 10°, as long as they're uniformly spaced. On the vertical scale, plot each voltage over each temperature reading, again the plot points must be uniformly spaced. Once you have all your points plotted, connect the dots with a straight segment between each one. Is the overall plot a straight line, a jagged, "random walk," or some recognizable shape? What conclusions can you draw?

Moving to your computer, if you have a spreadsheet program, like *Excel* or similar, you can do this a little more elegantly. Simply type each data point into a cell in a column. Once you have all the data in a column, select the entire column (by clicking on the column heading) and choose INSERT/LINE CHART. This will insert a nice formatted graph of your data into the spreadsheet. You can then edit the chart and label the axes any way you want. *Excel* took all the best features of some fine, high-end scientific graphics programs and made them available for us normal human beings. We'll be using spreadsheet programs a lot in later chapters. For those of you who don't want to use Microsoft, there is also *LibreOffice* (**www.libreoffice.org**), a free office suite that has a completely compatible spreadsheet program. Using a spreadsheet program saves graph paper (in theory — you'll probably want to print out your nice graph in three or four different full-color versions).

Of course, there are degrees of computerization available for the lab. In the above example, we used the computer to analyze the data, although not very thoroughly, but we still had to insert the information by hand. Wouldn't it be nice if we didn't have to read the DMM, but rather let the computer take each voltage measurement? Indeed it would, and this is where we enter the wonderful world of automatic data acquisition.

AUTOMATIC DAQ

In order for a computer to do any "number crunching" we have to provide data in a form it understands, which generally means binary numbers, 1s and 0s. The process of taking a sample of some physical value (say, voltage) that can come in an infinite number of values and converting it into a string of 1s and 0s is known as analog-to-digital (A/D) conversion.

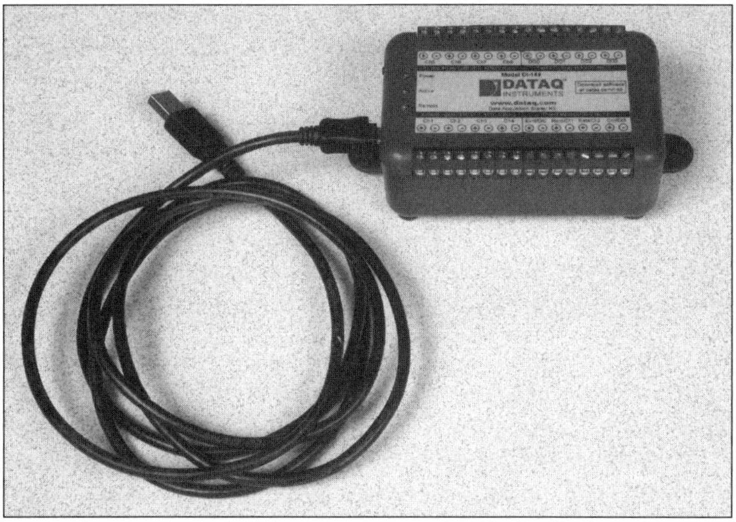

In our saltwater demo, the DMM performs the A/D conversion for us. It takes a *continuum* of voltages and converts them into a *finite* number. The display is in a digital form and quite accurate. However, we still had to

Figure 4.4 — Low-cost data acquisition (DAQ) modules, such as DATAQ Instruments' DI-149 pictured above, help make radio science affordable and accessible to any interested radio amateur. [Thomas Griffith, WL7HP, Silver Impressions, photo]

read those digital numbers and type them into the spreadsheet, which can be slow and tedious. Enter the DAQ device (**Figure 4.4**) into our ongoing demonstration.

Demo 4.3 — Still More Materials and Procedures

- DAQ card or module with software
- Twisted pair wire
- Thermocouple

The handy device is a computer card or outboard hardware module that converts analog voltages into numbers that the computer understands. DAQ hardware prices can range from around $50 to several thousand, depending on how quickly and accurately it does this conversion process, as well as how many channels of information it can process. The least expensive DAQ hardware plugs into the USB serial port of your computer, just like a card reader or similar peripheral. You can still get DAQ modules that plug into a standard serial port, and there are actually some performance advantages of this ancient technology.

Your DAQ device will almost certainly come with some software to control its behavior, such as how often it takes a sample, what voltage range it will be looking for, and the like. A piece of DAQ hardware that is designed to take samples at very low rates (for example, every few seconds) is known as a data logger or data recorder and is the easiest way to get started with this automatic DAQ business. High-performance DAQ devices can take up to millions of samples per second, and you will pay accordingly. The software package that comes with your DAQ will probably also have some nice graphing or other analysis software thrown in. Make sure you read *all* the supplied instructions — there are a lot of useful extra features you probably won't run across by trial and error.

APPLYING AUTO DAQ TO OUR DEMO

Returning once again to our experiment, let's say you've got a four-channel data logger connected to your USB port, and all the software is installed. We'll use Channel 1 of the data logger to start with. Since right now we're looking for voltage, we can remove the DMM probes from the fixed resistor, and connect the DAQ device to the resistor with twisted pair wire. (Although you can do this with alligator leads, this practice can be very susceptible to noise and other ills, so it's good to get into the habit of using appropriate connection methods. Your DAQ documentation will have all kinds of information and hints on how to do this properly, depending on the particular application).

Start taking voltage measurements immediately with our computer and DAQ. Your software will have a window to show a running tab of each channel's input voltage, say one number per second. If Channel 1 shows numbers around 10 V, and all the other channels are 0s, it's safe to say that you have it wired up right.

Well, we still have three more channels we can play with. Wouldn't it be nice if we could connect our DAQ device to the candy thermometer as well, so we can automatically record the temperature of the salt water? It sure would be, but we'd probably have to use something other than the candy thermometer. The *thermocouple* is the ideal choice. A thermocouple consists of a pair of dissimilar metals welded together, and when you heat up the junction, a voltage appears across the welded joint. (By the way, it works backwards as well, meaning if you apply an external current to a thermocouple, it gets cold. The Peltier, or *solid-state refrigerator*, works on this principle. Dunking a thermocouple into the solution will also give you a pretty accurate temperature reading.)

Now, the voltage you get from a thermocouple is something you can feed into the DAQ device's Channel 2. The DAQ will digitize this voltage and display it on the Channel 2 column. With the right gymnastics, we can calibrate this number as a temperature.

Why gymnastics, you ask? Well, because the thermocouple itself is not a perfectly linear device. The voltage you get is not exactly proportional to the temperature. Most of the time, it's pretty good, and for our experiment, most thermocouples are more than adequately linear. (For precision measurement, you want to use a *conditioning amplifier*, which is a device you put between the thermocouple and the DAQ device to compensate for the temperature versus voltage properties of the thermocouple. A conditioning amplifier is interesting in itself, because it generally takes the form of an *analog computer* or *logarithmic amplifier*, which we'll meet again in the next chapter.)

Returning to our handy graphing software, we can instruct it to plot Channel 2 data on the X axis and Channel 1 data on the Y axis, and then watch the data being plotted automatically in real time. No more tedious thermometer watching or DMM reading. Pretty slick!

Demo 4.4 — Final Tweaks to Materials and Procedures

Being thrifty sorts of radio experimenters who don't like paying for stuff we aren't using, we can't help but notice that we still have two unused channels on our DAQ device. What else could we put in there that might have an effect on the saltwater resistance? For the hopelessly curiosity driven (though the price tags may leave the thrifty a bit pale), we'll now add:

- Capacitive pressure transducer
- Capacitance bridge

Well, as you'll recall from our earlier discussion on conditions, both temperature and pressure affect water density. We've already dealt with temperature already, so let's consider barometric pressure. Would it have any measurable effect on the saltwater conductivity as well as density? Take a guess and then test it out.

So, how do we feed barometric data into our DAQ widget? Actually, there are several ways to do this. One of the most accurate methods is with a *capacitive pressure transducer* (available online). *Capacitance* is the ability for two conductors in close proximity to maintain an electrical charge between them. Ben Franklin used a primitive form of capacitor, called a Leyden jar, in his famous lightning experiments. The capacitive pressure transducer method uses a sealed diaphragm with a conductive metal plate on either side of the diaphragm. The diaphragm squishes down a bit when the barometric pressure increases, bringing the plates closer together.

The capacitor plates in the barometric transducer are fed a small high-frequency signal and then measured with a *capacitance bridge* (easily constructed), which gives an output voltage proportional to the capacitance. This voltage, which is proportional to barometric pressure, can then be fed to the DAQ channel. True, you probably will see no connection between barometric pressure and the saltwater resistance, but you'll have a long-running record of barometric pressure anyway, which may be useful for some *other* experiment down the road.

We still have one input channel dangling in the breeze. What other piece of relevant information can we apply? My personal preference would be power supply voltage. Although our supply is presumably stable and accurate, we should, by now, be highly averse to taking anything for granted. So, finally, run a twisted pair of wires right from the power supply output terminals to the Channel 4 input terminals of the DAQ widget.

There you have it. You've run fully automatic DAQ of two highly relevant channels of data, one probably irrelevant channel of data, and one semi-relevant channel of data. This is lab science at its best. Or maybe, just at its most automated.

Software Tools for Hard Numbers

Ever since man started counting things, he's been looking for easier ways to do the job. The abacus (**Figure 5.1**) is one of the earliest calculators still in use, but it was by no means the first. The word calculate comes from the root *calx*, a small pebble used in counting. Things have gotten somewhat more complicated.

From time immemorial, people have found it easier to count things than to measure them. As a race, we seem to be uncomfortable with things that come in indefinite quantities, although the very nature of the universe

Figure 5.1 — The abacus is one of the world's first "digital computers" and is still in use in many parts of Asia. [Photo courtesy Thomas Griffith, WL7HP, Silver Impressions]

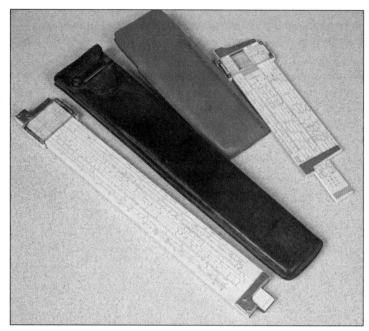

Figure 5.2 — The slide rule, now a collectors' item, was until recently standard equipment in every engineer's arsenal. [Photo courtesy Thomas Griffith, WL7HP, Silver Impressions]

is based on a continuum of values — at least on easily observable scales. We like to chop things up into discrete units that we can count.

This need to quantize everything has not always been universal, however. Lots of engineers cut their calculating teeth on a slide rule (**Figure 5.2**). The slide rule is a good reminder of the logarithmic nature of the universe. In fact, the slide rule performs all its calculations by addition and subtraction of logarithms. I encourage everyone to learn how to use a slide rule, not as a substitute for modern calculators, but because of the wonderful insights it gives to a lot of mathematical concepts and physical concepts.

OP AMPS AND LOG AMPS

Although the slide rule has fallen into general disuse, the electronic equivalent of the slide rule, the analog computer certainly has not. The analog computer is an electrical circuit that performs mathematical operations directly. It works by virtue of the fact that all electrical circuits follow strict mathematical laws. The analog computer is unique in that measurement and calculation are not separate functions, but are one and the same. Analog computers function in real time, their speed limited only by the speed of electricity itself.

The ubiquitous *operational amplifier* (op amp), which is a building block in everything from audio amplifiers to electroencephalograms, was originally developed to perform mathematical operations, hence its name. The coveted extremely high gain and bandwidth of modern op amps is actually a byproduct of their original purpose. Linear op amps can perform precision addition and subtraction functions. However, the more interesting variation of the op amp, the *logarithmic amplifier*, or log amp, is capable of performing nearly any mathematical function. And it performs these functions in the same manner as a slide rule: by the addition and subtraction of logarithms.

Log amps are extremely important components of many data acquisition systems, because they greatly increase the dynamic range of the signals that can be accommodated. They are integral components of any modern Amateur Radio receiver as well. Log amps also can be used to create very complex waveshapes, and are integral parts of many types of function generators.

So we see that the terms *computer* and *digital* are not synonymous — not all computers are digital. The effective radio scientist should be able to move between the analog and digital realms with ease, and understand the advantages of each approach to problem solving.

FROM COUNTING TO COMPUTING

The personal computer has become so pervasive that it's easy to forget its original purpose: to compute. Most computer users don't think much about the actual number-crunching that goes on inside a their machines, but as a scientist you'll need to, at least a little bit. You should also get into the habit of good computer methods right from the beginning; it will pay big dividends later. This requires knowing a little bit about computer math. Even if you don't do any actual programming yourself, many scientific applications can be "hot-rodded" if you know what's under the hood.

Although computers are screaming fast compared to just a few short years ago, the demands of modern scientific applications have followed right on the heels of the increased performance, and nothing requires as much computer brainpower as modern *computer modeling*. This software simulation function has made its way into every industry from petroleum refining to genetic engineering, and it will be an important topic for us in this book.

ASCII NOT FOR EVERYTHING

Probably nothing has contributed more to bringing the computer into the every home than the ASCII (pronounced "ask E") standard — and probably nothing has contributed more to "computer cholesterol" than the ASCII standard. It's definitely a two-edged sword.

ASCII stands for American Standard Code for Information Interchange. It is an alphanumeric code that allows us to create text files that contain both letters and numbers. ASCII is capable of handling 128 symbols using 8 bits of binary information (the eighth bit is used for parity in some systems). Each symbol in ASCII uses 8 bits, whether it's a letter, a numeral, a punctuation mark, or some special control, most of which are now obsolete. It's a great system for some things, such as sending text files containing letters and numbers back and forth between computers or over the Internet. It is *not* a great system for number crunching. In order for a computer to do any actual calculations, it has to locate *numerals* in an ASCII stream, convert them to a numerically useful binary form, perform the binary calculation, and then convert them back to ASCII, so we can see the results on a computer screen or other readable form. This is not an efficient process for high-speed scientific applications, as you will see.

Modern data entry systems require ASCII, whether that system is merely typing 1,2,3,4,5,6,7,8,9,0 in a word processor like this, or entering those numbers in a spreadsheet. However, the sooner we can get that number into a binary form, the better life will be. Not only does numerical data stored in binary form take up much less hard drive space than ASCII data, it's also faster to extract (and write) binary data from or to a hard drive (of any type) than ASCII.

JUST MY TYPE

Nothing will increase the performance of a computer more than careful attention to data *type*. What do we mean by that? Type is simply the number of "decimal" places we use to perform a binary operation (I use "decimal" in quotes, because binary math by definition is not decimal — it's *binary*; just keep that in mind for the discussion that follows). There are only two possible numbers for any decimal place: a 1 or a 0.

Strictly speaking, ASCII is not a data type, but it can be used as an example of a bad way to handle numerical data. ASCII always uses eight decimal places (bits) to express anything from a letter of the alphabet to a left parenthesis. It uses the same number of bits no matter what. This makes the code very flexible, but very inefficient. As a result of this flexibility many systems store data in 8-bit bytes, whether the data is really ASCII or not.

Let's revisit our saltwater experiment to drive home this point. Say we aren't really interested in the precise static temperature of the solution, but only when the temperature goes above, say, 100°F. Now say we've programmed our DAQ device to take a sample every second and give us a 1 if the solution is above 100° and a 0 if it's below. (It's an easy job to write a scrap of computer code to do this "go/no go" function). Every second then, depending on the temperature, we write either a 0 or a 1 to a file on a hard drive. However, these numbers are going to be communicated and stored as if they were in ASCII. A 0 stored this way is 00000000, and a 1 is 00000001. So every second, we use up 8 bits of storage space. No biggie, you say; you'll be retired before you ever use up your hard drive space. But what if you were to take a million samples a second? How about 24 channels of data sampled at a million times a second? You'd go through a gigabyte of hard drive space every 41.6 seconds. You'd grind through a terrabyte in 11 hours. That's not enough to run even a weekend experiment.

So how big a data type do we really need to express a 0 or a 1? How about just a single bit? Why use 8 bits to express a piece of data that can be handled with 1? A 1-bit data type is commonly called Boolean. Most scientific programs can handle Boolean data just fine, and store it as such. Granted, you're not likely to be sampling such rudimentary data as in the above experiment at 24 megasamples/second. But lesser "crimes against type" are committed daily by careless computer users and that means a loss of performance. Choose your type carefully; one size does *not* fit all.

FLOPPING AROUND

One of the more common data types you'll encounter in scientific computing is called floating point. High-speed computers are generally rated in the number of gigaFLOPs (billion FLoating point OPerations) or teraFLOPs (trillion floating point operations) they can perform per second (actually FLOP speed is much more meaningful than clock speed for any computer). TeraFLOP computers used to be the exclusive domain of supercomputers like the Cray, but recently, *cluster computers*, using massive stacks of off-the-shelf machines, can nearly match these speeds. You may not encounter a true supercomputer any time in your career, but it's quite likely you'll encounter a cluster computer system at some point.

No matter what field you're in, scientific data can range from extremely large values to extremely small values. Thinking decimally for a moment, so as to not be confused by the binary conversion process, let's look at a "type" (again in quotes, because we wouldn't really do this decimally) that can handle a number ranging from 1 million all the way down to 1 millionth (a data range of 1000000 to 0.000001). Here we have six places to the right of the decimal and six places to the left, for 12 decimal places total. We could create a type with 12 places or bits. But we would seldom need all those bits. There would almost never be a need to express the number 1000000.000001, for example. However, we might occasionally need to express 1000001 or 0.1000001. Both of those numbers require the same amount of precision: seven decimal places, or "bits."

How can we allow the same sized type to give us the same precision regardless of whether the number is huge or tiny? By allowing us to slip the decimal place around *after* we do the calculations. This is what floating point math is. You simply do the operation in two steps; the first one disregards the place value entirely, and the second one shifts it to where it needs to be. You do need a few bits to tell you where to poke the decimal point in so there is a small price to pay in memory usage, but it's negligible considering the overall efficiency. Most common floating point math is based on a 32-bit number. This gives you a vast range of values, and only a small loss in precision over fixed point operations.

ROLLING YOUR OWN

Despite the vast amount of scientific software available for the lab, sooner or later you will probably have to write some of your own software, or at least modify someone else's handiwork. If you're doing any kind of Research and Development (R&D) work, you will eventually go where no software has gone before. This can be a source of great dread or great opportunity.

There are hundreds of programming languages out there, and they each have their strong and weak points. Of course, if you're modifying an existing piece of software (most likely written by a colleague who left you with no documentation) you will have to use the language in which it was originally written.

A good amount of in-house scientific software was written in C, C++, or Fortran. Many people consider Fortran an obsolete language, but it is still the core of most numerically intensive programming, as well as modeling software, and you can never go wrong by learning it. Contrary to what you might have heard, Fortran will be around for a long time. You'll find it behind *MATLAB* and the free-to-download *Octave* numerical computation software and *Numerical Electromagnetics Code* (*NEC*), which is the basis of all modern antenna modeling systems, among many others.

Let me camp out for a bit on one of the above programs: *MATLAB*.

If you are in the sciences, you *will* end up using this software from MathWorks (**www.mathworks.com**) at one time or another. *MATLAB* is the premiere modeling and imaging program, and it is based on a very simple premise: all large collections of data can be arranged in the form of a matrix (the name is an abbreviation for Matrix Laboratory). A matrix, of course, is simply an array of numbers in a certain number of dimensions, arranged in rows, columns, or even heaps.

MATLAB uses *linear algebra* to manipulate and display any imaginable collection of scientific data in an astonishing variety of graphic representations. You can take 3D (or more than 3D) objects and rotate them around any axis at will, turn them inside out, revolve them, or morph them into other objects. There are dozens of "toolboxes" within the program that let you model any type of system, such as chemical, physical, mechanical, and electronic. I want to reiterate, however, that a computer model is not a substitute for knowledge, experience, or an actual physical experiment. You need to test computer models against something with known behavior, just as you need to calibrate a thermometer against a known standard.

Another must-have program for any serious radio science investigator is *Origin*, published by OriginLab (**www.originlab.com**). *Origin* is probably the very first program I ever used of any kind, and it's also the tool I used to produce many of the cool graphics in this book. The earliest versions of the software simply created a line graph from spreadsheet data you entered, but the latest version can produce a staggering variety of 3D graphs and offers number crunching capacity suitable for just about any imaginable scientific discipline. *Origin* is also one of the most user-friendly of all scientific number-crunching programs.

We'll be returning to both *MATLAB* and *Origin* later in this book.

Optics — Where Radio Sees the Light

Optics can be thought of as the first cousin of electronics, because both disciplines work exclusively with *electromagnetism*. Visible light is made of the same stuff as radio, just on a scale many orders of magnitude smaller than "conventional" radio waves and electric currents (remember, an order of magnitude is a 10:1 ratio of difference between two phenomena). Radio waves are generally measured in meters, whereas light is measured in Ångströms, which are equivalent to one ten-billionth of a meter. Measurements, therefore, have to be about 10 billion times as fine at visible light frequencies as they do at normal radio frequencies. If your interests lead you toward the study of optics, take heart; this is not as difficult as it sounds. As with any scientific investigation, you simply need to pay close attention to detail.

By studying optics we can learn a lot about the things that affect us radio amateurs. To a first order approximation, radio propagation follows optical principles, at least most of the time. Among the many common optics principles that are directly relatable to radio we find: reflection, refraction, diffraction, dispersion, interference, polarization, collimation, free space attenuation, and absorption. There are also some phenomena that are almost directly relatable between optics and radio, such as sporadic E (E_s) propagation, pseudo-Brewster angle, and birefringence (which we'll touch on in Chapter 12).

RUNNING INTERFERENCE

One of the most common, and demanding, tests of optics precision is in the field of *laser interferometry*, which uses superimposed electromagnetic waves to extract information about physical objects. Being able to perform laser interferometry is a very good test of your laboratory skills and methods. If you can do this reliably and consistently, you can probably do anything else in the optics lab setting. Let's look at a real-world example first.

Thanks to the consistency in the universe, we can take a phenomenon as common as waves on the surface of water and use them to precisely describe properties of electromagnetic waves, such as light. One of those properties is *wave interference*. Say you drop two rocks into a pond a certain distance from each other; the ripples they cause will radiate from the "splash points" in concentric circles. When the two expanding rings of ripples overlap each other, the phenomenon of interference will result. At some locations the heights of the waves will add; at others they will subtract, or cancel out. Where the ripples add we're observing what's known as *constructive interference*, and where they cancel out we see *destructive interference*. The destructive interference points are very well-defined locations (where the surface of the pond is perfectly calm), and if you know the speed that the waves travel across the water, you can very accurately determine the distance between the two original splashes by looking at the destructive interference points.

This principle is used in optics to obtain very precise measurements of distance (and movement, with a small variation in setup). We can measure the distance between two different points by looking at the interference patterns of light reflected by those points, determining it to within the wavelength of the light we're using (within nanometers).

APPARATUS OF OPTICS

Other than the numbers involved, light and radio waves are actually the same thing. Many optical instruments are nearly identical (in nature, if not in size) to those used for normal radio frequencies, while some are quite unique. Because the wavelength of light is much smaller than normal radio frequencies, optical instruments can be much smaller than their equivalent radio frequency hardware. We mentioned that light is generally measured in *nanometers,* whereas, radio frequencies can be *thousands* of meters long. Even *microwave* frequencies have wavelengths measured in centimeters or millimeters, making them several orders of magnitude longer than visible light.

Optical experimentation generally takes place on an optics bench. This will be an extremely rigid and stable bench or table upon which you can mount various lenses, prisms, diffraction gratings, and even lasers. Professional optics benches have an array of threaded holes into which you can screw supporting posts for the various optical instruments. All optics hardware is painted flat black to reduce reflections from undesired

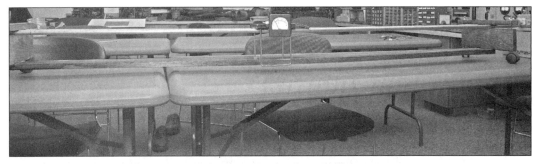

Figure 6.1 — Three views of the Lecher Wire, one of the oldest instruments of radio science. The Lecher Wire measures wavelength of radio signals directly and offers insight into a number of radio phenomena, such as standing waves on transmission lines. [Thomas Griffith, WL7HP, Silver Impressions, photos]

light sources. Sometimes the walls of the room are painted flat black as well, or covered with black drapes. With the increasing prevalence of lasers, this presents an additional safety factor, as reflections of laser light from metallic surfaces, or even glossy painted walls, can present some danger. Some optics labs have seismically isolated floors, so that vibrations from distant earthquakes, or even passing trucks, don't get transferred to the bench and upset measurements.

One of the oldest instruments we use in radio is a device called the Lecher Wire, or

Lecher Line (**Figure 6.1**), which measures the wavelength of radio waves directly. Like the optical interferometer, which measures objects using the interference pattern of two waves of light, the Lecher Wire uses the principle of interference. Even in its crudest form, it is capable of precise readings, and it's also extremely easy to build, consisting of nothing more than two parallel copper wires or rigid tubes stretched across some kind of supporting structure. Here's how it works: A radio frequency voltage detector is slid along the Lecher Line to detect the voltage at any particular location. If you feed a source of radio frequency into one end of the Lecher Line (typically in the VHF or UHF frequency range), the RF detector will show varying voltages as you move the detector along the line. This is caused by interference between the forward-going wave, and reflection from the open end of the Lecher wire. You can also put a short circuit on the far end of the Lecher Wire, which will move the RF voltage maxima and minima by one quarter of a wavelength. Not only is the Lecher Line useful for measuring wavelength by measuring the distance between the minima, but it is also capable of demonstrating a wide variety of *transmission line* phenomena.

The first practical method of observing wave interference was developed by Albert Michelson in the late 1800s, and the Michelson interferometer is still the basis of all modern laser interferometry. Today interferometry commonly uses a helium neon (HeNe) laser for the task, an innovation that gave the first commonly available *continuous beam laser*, as opposed to the *pulsed laser*. While fairly inexpensive as lab lasers go, you can also build an interferometer with a laser pointer, which uses a somewhat less stable laser. For demonstration purposes, though, the laser pointer interferometer is more than suitable. Of course, Albert Michelson didn't have lasers in his day, so he used the next best thing: the monochromatic (one color) light from a prism.

Most real-world radio propagation used by terrestrial radio amateurs involves a combination of reflection and refraction. The ionosphere (as well as other lower layers of the atmosphere) is more akin to a prism than to a mirror. But even this is a bit inaccurate: Actually the property that bends different colors of light at different angles is *dispersion,* not *refraction.* Dispersion is a subset of refraction, but refraction can be dispersive or non-dispersive. Telescope lenses, for instance, work by refraction. You *hope* a telescope is not dispersive, and you generally take great pains to assure that it's not, such as by means of *achromatic* lenses and such. One of the primary advantages of a *reflector* telescope over a *refractor* telescope is that a mirror is inherently non-dispersive.

RADIO THROUGH THE LOOKING GLASS

Segueing this to radio, when it comes to high-gain radio antennas, most radio amateurs visualize something like a parabolic dish. Although the parabolic dish is not the only possible configuration for a radio antenna, it's probably the most obvious solution. Most of us know all about reflector telescopes and refractor telescopes, which use parabolic reflectors and refractors. But when it comes to radio antennas, we don't usually think of refractor antennas. One may be forgiven for even wondering if there is such a thing as a refractor antenna. Indeed there is, but they are only truly practical at microwave frequencies. And even as such, they are not terribly common. But a radio lens can indeed be built out of any dielectric material, Teflon being one of the most common. A Teflon lens at microwave frequencies works in precisely the same fashion as a glass lens at optical frequencies.

But even more interestingly, a radio lens can be built out of a plasma (more on plasma later).

One can actually create a lens-shaped (or more commonly, a spherically shaped) region of higher or lower electron density inside a plasma chamber, or even the ionosphere itself. Admittedly these structures are somewhat short-lived, but not so fleeting as to prevent some really interesting experimentation. At HIPAS (HIgh Power Auroral Stimulation) Observatory in Alaska, where I worked for many years, we were able to create plasma lenses in the ionosphere that very dramatically increased the effective radiated power (ERP) of radio signals passing through them! We used a "pump" frequency of one value to "heat" the region to make the lens, and then passed another frequency through the lens, using remote receivers looking at the signal reflected from a much higher altitude to determine the signal strength. Pretty cool stuff. Not necessarily *very useful* stuff, but cool nevertheless.

WHERE THINGS DIVERGE

Because so many aspects of radio propagation are described by optics phenomena, it's in exploring the ways radio propagation *departs* from expected optical behavior that we can learn even more about the ionosphere and other related matters. Knowing what something *isn't* is often as useful as knowing what it *is*.

Let's look at sporadic E propagation (E_S) as an example of something that has particularly optical behavior. There are a lot of things we *don't* know about E_S, but there are some things we can quite confidently say about the phenomenon. For instance, we know that E_S reflections are nearly *specular* (mirror-like), which means there is a sudden transition between no free electrons and a whole lot of free electrons in a very short vertical distance. E_S reflections are also *frequency independent*, unlike F-layer reflections, further confirming the very sudden transition from a *non-conducting* region to a highly conductive region. Now, just where these clumps of electrons come from in the first place is still the subject of a lot of speculation, but there is no doubt that we do have clumps of electrons. Ablation of micro-meteors is one probable source of E-layer electron clumps, as is jet stream "wind wiping." Using both radar methods and conventional triangulation, we can quite accurately pinpoint the location of these electron clumps. This is because electron clumps create refraction clumps, and refraction is a well-defined optical property.

Despite the similarities between light and radio, at times the analogy falls apart upon examination. At normal radio frequencies, we can pretty much ignore the *quantum* properties of matter. We don't usually speak of photons when dealing with radio antennas and such. It's not that there *aren't* radio frequency photons; it's just that the energy levels involved in radio are insignificant on the atomic level.

At visible light frequencies, the "antennas" involved are atoms themselves. Wavelengths are on the same scale as atoms or molecules. Only at the highest humanly achievable radio powers can we even excite an atom to a higher energy state in the ionosphere. At HIPAS Observatory and some similar facilities, *airglow* (a faint luminescence of the upper atmosphere caused by air molecules and atoms temporarily attaining an excited state) was achieved using HF pump frequencies. But it took "everything they had," and the results were only visible with extremely sensitive optical methods. Even a gigawatt of ERP, which is what HAARP produces, doesn't even remotely approach energy levels sufficient to actually *ionize* a single atom.

THE PSEUDO-BREWSTER ANGLE

Another peculiarly optical-like phenomenon is the pseudo-Brewster angle. We know that only vertically polarized radio signals can propagate along the surface of the Earth. If you try to propagate a horizontal wave along the surface, the E field (horizontal component) is shorted out. You can't create a difference of potential across a short circuit. (Since the Earth is *not* a dead short, this phenomenon is modified a bit, hence the "pseudo" in pseudo-Brewster). In reality, vertically polarized waves are refracted *into* the Earth's surface, but since the surface is curved, they tend to follow the contours of the planet. This is what allows a ground wave to propagate well beyond the line of sight.

If you want to demonstrate a "true Brewster" phenomenon, you can do so with a pair of polarized sunglasses or polarized filter and a microscope slide. Set the microscope slide flat on a bench in view of a light bulb. The light bulb needs to be overhead, but at a distance that puts the reflection at a low angle. Look at the slide to see the reflection of the light bulb on the surface. Now look at the reflection through the polarized filter. As you rotate the filter, the reflection of the bulb will completely disappear at a particular angle of rotation. This demonstrates that the reflection from the glass slide is polarized.

This is just one example of how we can apply optics to Amateur Radio and vice-versa.

7 The Electromagnetic Spectrum at a Glance

Radio amateurs are allocated valuable portions of radio frequencies throughout the electromagnetic spectrum. This is a vast privilege and a vast responsibility. When compared to the other users of the radio spectrum, such as commercial, military, and scientific entities, the total spectrum space is pretty minuscule, but the wide variety of frequencies we have is quite astonishing. In the United States, we have frequency allocations ranging from medium wave clear up to millimeter wavelengths. Although hams are quick to complain about the lack of elbowroom on the amateur bands, especially when propagation is good, we also have vast tracts of unoccupied "RF desert land."

This is especially true in our microwave bands, of which we have 12, almost totally unused. Although these are often deemed to be of little practical value for amateur communications, they hold immense value for scientific investigation. In fact, each of these 12 bands exhibits sufficiently different properties that we should not risk losing any of them for lack of use, at least if we're as interested in "advancing the radio art" as we claim.

At the other end of the radio spectrum, the "subterranean" hinterlands of radio, we have no restrictions, at least as far as reception goes. A fascinating, bizarre world of "natural radio" exists down below about 22 kHz, where any radio aficionado, licensed or not, may quite literally eavesdrop on "the music of the spheres." Here one finds a diverse array of oddities like whistlers, choruses, auroral roar, auroral hiss, and a few signals nobody even has a name for yet. These phenomena are all ionospheric in origin (at least to the best of our knowledge), but we might find we've been barking up the wrong tree all along. This is still largely unexplored and unexplained territory.

Figure 7.1 — Sometimes it seems that radio is nothing but winding a lot of coils. This is partially true. [Thomas Griffith, WL7HP, Silver Impressions, photo]

We old-time radio amateurs are fond of saying that radio is so simple "a caveman can do it," a statement that masks the subtlety and sophistication of modern radio techniques. Indeed, early radio equipment can now seem a little "prehistoric" and often amounted to little more than coils of wire (**Figure 7.1**). Back then, experimenters paid little heed to spectrum use, or even the possibility that there might be entities other than ourselves who might be interested in using the precious resource known as radio spectrum. The spark gap transmitter was state of the art at one point in time, and a signal from such a technological monstrosity generally occupied most frequencies of the known radio spectrum *simultaneously*. Fortunately, the spark gap transmitter is today little more than a laboratory or museum curiosity, but it's also a useful reminder of just how easy it is to generate a radio signal.

In between the extremes of spectrum is HF radio, which is where we are likely to see the widest variety of both linear and non-linear ionospheric phenomena. We've seen these terms before in this book and we will visit them again. They both have mathematical significance, but it's probably helpful now to add the more intuitive definition of these terms.

In plain English, a linear system is one in which there is a direct connection (as well as an immediate one, generally speaking) between cause and effect. If you push Button A, you get Response B, each and every time. Human beings are generally very comfortable with linear systems, because they give the illusion that humans have some predictable control over their universe. Linear systems are generally *proportional* as well as predictable. When you push an object harder, you expect it to move faster, for instance. You also reasonably expect the object to move in the same direction as you push it. Probably the majority of what we observe in typical Amateur Radio activity falls well into the linear system category. As we will learn in subsequent chapters, however, it is when radio exhibits *non-linear* behavior that we have the most exciting prospects for discovery. Some of what we will observe is beyond bizarre. But in order to understand the bizarre, it helps to understand the *normal* first.

TOURING NORMAL

Before we go too far down the road to bizarre, let's complete our tour of the electromagnetic spectrum. We should note that there are a lot of pretty arbitrary lines of demarcation within the electromagnetic spectrum. There are some fairly logical historical reasons for the electromagnetic spectrum being subdivided into HF, VHF, UHF regions, etc, though they really don't have any solid physical basis. Radio doesn't suddenly behave differently when you move from the VHF range to the UHF range, for instance. There is a lot of overlap between these categories, but they are fairly useful divisions in everyday practice.

Although most of us in the United States have only recently accepted the metric system, and that only with a lot of kicking and screaming and gnashing of teeth, radio measurements have always been metric. Before anyone knew how to measure radio frequencies, radio signals were measured by wavelength in meters. It also turns out that the speed of light (or any other form of electromagnetic radiation) is very close to a round metric figure (300 million meters per second), which is a lot closer than any round English system measurement comes.

The known radio spectrum is conveniently subdivided into orders of magnitude of wavelength, which gives us the following general pattern:
- MW (mediumwave): 1000–100 meters
- SW (shortwave or HF): 100–10 meters
- VHF (very high frequency): 10–1 meters
- UHF (ultra-high frequency): 1–0.1 meters
- EHF (extremely high frequency): 0.1–0.01 meters
- SHF (super high frequency or millimeter wave): 0.01–0.001 meters

Various government agencies have scooted these definitions around at times; for instance, the US military services place the dividing line between VHF and UHF at 200 MHz, which is actually 1.5 meters, not 1 meter. Also, the traditional divisions on the low frequency end, such as between VLF and ULF aren't quite as tidy. But for the most part, these divisions are quite useful for describing the general behavior of different radio bands.

Not only are the demarcations between the various bands somewhat flexible, but since there is a 10:1 ratio of wavelength between the top and the bottom of each of these ranges, one might expect there to be a notable difference in behavior between radio signals within any of these bands — and one would be correct. The difference between radio propagation at the high frequency extreme of the HF band and that at the bottom end of the same band is as different as day and night, quite literally.

The 500 kc Lowdown

No overview of radio science would be complete without mentioning the 500 kc community. Full details of this organization can be obtained on **www.500kc.com**. A few years ago, the ARRL obtained permission from the FCC to allocate a limited number of experimental stations near 500 kHz (600 meters). This frequency has an interesting history. For about 100 years, it was the international maritime distress frequency. Until quite recently, all large sea-going vessels were required to keep *radio watch* on 500 kHz for ships in distress. This frequency range afforded reliable worldwide coverage day and night, due to the effective ground wave propagation over seawater.

With the advent of GPS and other more advanced navigational and distress communications systems, however, 500 kHz was re-allocated for experimental and developmental work. This has become an effective proving ground for weak signal processing and less-than-optimal-sized antenna experimentation. It's a great way to develop scientific radio skills, particularly weak signal reception methods that can be applied to any part of the radio spectrum. Time-honored weak signal detection methods such as *synchronous detection, coherent detection,* and *lock-in amplifier* technologies are employed by a number of radio amateurs on 500 kHz, as well as state-of-the-art DSP methods.

NOW, A LITTLE LESS THAN NORMAL

Did you know that you can engage in extraterrestrial communications with any class of Amateur Radio license, or no license at all. With the exception of satellite communications, and perhaps moonbounce, amateur extraterrestrial communications will be receive-only, which means no license necessary. Of course, the oldest form of extra-terrestrial radio science is astronomy. The human eye is the only organ equipped to detect electromagnetic radiation (at least to the best of our knowledge), which is why optical astronomy is the oldest form of radio science.

It's no accident that many of the most dedicated Amateur Radio experimenters are also amateur astronomers. Until fairly recently, nearly all of what we knew of the universe outside of our immediate neighborhood was based on what we could see. We've generally assumed that most of what was important in the universe was hot enough to emit visible light. There's increasing evidence, however, that the universe is composed largely of dark matter (by "dark" we don't necessarily mean stone cold, just not hot enough to emit visible light). However, the universe radiates everything "from dc to daylight" and there's a lot of radio out there.

Radio astronomy techniques that used to be out of the reach of all but the most well-heeled hams are now available to nearly everyone, with the ready availability of surplus satellite dishes and low-noise amplifiers with performance only dreamed of just a couple of decades ago. Coupled with modern digital signal processing (DSP) software, nearly any ham can do world-class radio astronomy.

Back on our planet, some of the most interesting and unexplored radio territory is to be found at extremely low frequencies (ELFs), below 200 Hz or so. Most submarine communications take place in this region, though there is some doubt as to whether this is actually radio propagation, as opposed to merely *magnetic coupling.* Because the wavelengths are on the order of thousands of kilometers, it's extremely difficult to measure the rate of attenuation, which would conclusively prove whether signals "down there" are really radio waves.

Needless to say, it's extremely difficult to generate powerful radio signals down at such long wavelengths, and it is largely to this end that ionospheric heating research facilities like HAARP (High Frequency Active Auroral Research Program; **www.haarp.alaska.edu**) and EISCAT (European Incoherent SCATter; **www.eiscat.com**) exist. One of the major goals of non-linear ionospheric experimentation is to find an effective means of generating ELF signals without the use of hundreds of miles of terrestrial wire. ELF frequencies are the only ones that can penetrate seawater to any degree and are, therefore, the only frequencies useful for submarine communications.

This is one area where the ham community, by offering lots of receivers, has a greater potential to answer some crucial questions than the tiny number of institutional ELF receiving facilities.

Figure 7.2 — The author displays an experimental ELF coil, consisting of over 70 pounds of copper wire sandwiched between sheets of plywood. [Thomas Griffith, WL7HP, Silver Impressions, photo]

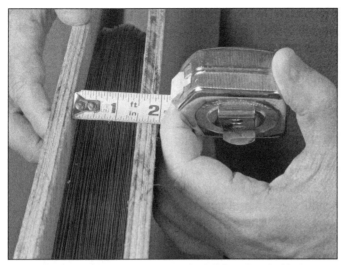

Figure 7.3 — This "ancient-technology" coil is interfaced with the high-tech, ultra-low-noise instrumentation amplifier of Figure 7.4 to create a sensitive ELF instrument. [Thomas Griffith, WL7HP, Silver Impressions, photo]

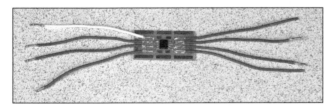

Figure 7.4 — In a classic example of old meeting new, coil and amplifier combine to seek out Schumann resonances (see text). [Thomas Griffith, WL7HP, Silver Impressions, photo]

Building effective ELF receivers is challenging, but it's not beyond the skill level of any dedicated radio amateur. We wryly refer to the total copper content (TCC) as the supreme measure of any radio station's performance. When it comes to ELF, this is particularly true — the more copper wire you can stretch out, or wind into a coil, the better your station will perform, whether for transmitting or receiving. In fact, the lower in frequency you go, the bigger your coils need to be. In **Figure 7.2** you see yours truly with an experimental ELF coil. The end goal of this is to be able to detect *Schumann resonances*, which are natural cavity mode resonances between the surface of the Earth and the lower ionosphere. They're on the order of 7½ Hz. Detecting Schumann resonances is a good test of one's ELF equipment. This particular coil has about 80 pounds of magnet wire, about 700 turns, and has nearly 1 H of inductance without a ferrite core (**Figure 7.3**). A special low-noise, high-gain op amp is used with the coil (**Figure 7.4**). As of this yet, we haven't been successful receiving Schumann resonances, but failed experiments are educational, too.

8 Free Space Radio

We've said that to comprehend the unusual, one needs to fully understand the normal, but perhaps the word *normal* should be replaced by the word *ideal*. Much of radio science involves exploring the abnormal or unexpected response of radio signal to various media. However, radio is simplest, and complies most closely with ideal theory, when in a perfect vacuum. For most of us humans, however, a perfect vacuum is not a normal state of existence, so we need to distinguish between ideal and normal in this regard.

In physics, there are many ideal concepts that don't exist in real life. One example in radio is the purely mathematical construct known as an *isotropic radiator*. By definition, an isotropic radiator is an antenna that radiates electromagnetic waves exactly equally in all directions. In reality, no such device actually exists. One can *approach* an isotropic radiator, but one can never build one, at least not with materials that occupy any volume of space whatsoever. Does this mean an isotropic radiator is meaningless? Not at all. It is an extremely useful tool for comparison, one that's commonly used as a yardstick against which to measure any real antenna. The simplest standard antenna we can actually build is a ½ wavelength (λ) dipole, which has a known gain of 2.13 dB over an isotropic radiator. There are indeed physical primary standard reference antennas; in fact, one can build one of these fairly simply.

Another closely related "impossible standard" in radio technology is the *inverse field measurement*, which describes a transmitting antenna's field strength *at the antenna*. Since it's impossible to measure the field strength at an antenna without occupying the same physical space as the antenna with one's measuring device, this presents a bit of a dilemma. So we have to rely on sound mathematics and accurate near-field measurements to tell us what the field strength really is at the antenna.

RAY WAYS

In free space — a perfect vacuum — a radio wave will have the smallest amount of attenuation possible. In a vacuum there is nothing there to absorb, reflect, convert into heat, or otherwise modify the behavior of a radio wave. If this is true then why *does* a radio wave in free space get weaker with distance? Or does it? (Don't take my word for it, but you'd probably suspect that it does, and there's a compelling body of evidence to confirm your suspicions.)

Free space attenuation, sometimes known as *geometric attenuation,* exists simply because we are taking a given amount of energy and spreading it out over a larger volume of space. To be perfectly accurate, it's actually the *power*, not the energy, that gets spread out. Power density of radio waves is measured in watts per volume (the actual units sometimes being a matter of preference).

For all practical purposes, radio antennas emit *spherical* waves. Regardless of the gain or pattern shape of the transmitting antenna, once you get a few wavelengths away from the source, the wavefront is essentially spherical in shape. The one notable exception to this is the so-called *collimated* beam, where all the rays are emitted in parallel.

We need to take a slight detour here to explain that there really is no such thing as an electromagnetic ray. Like the isotropic radiator, the ray is a mathematical construct to help illustrate what we can observe. In fact, no quantity of electromagnetic radiation can be confined to a finite number of rays. A ray is simply a *locus* of the point in space where some set of conditions is satisfied, namely intensity, polarization and direction, similar to isobars on a weather map.

However, the ray is a useful illustration of what happens in this discussion. If all the rays emerging from an antenna are parallel, there will be no free space attenuation. In reality, the closest thing we have to a collimated wavefront is what we get from a laser. In a laser, there are untold trillions of fundamental dipole radiators (atoms) that emit their portions of the radiation in phase. Obtaining the same degree of collimation that a laser has, at HF radio frequencies, would require an array of antennas hundreds of thousands of miles across. Needless to say, this is a somewhat impractical arrangement. Instead, we need to consider every radio antenna, no matter how elaborate, as being a mere "point source" of radiation, at least at meaningful distances from the source.

If all practical antennas are essentially point sources, one might be tempted to ask why build elaborate high-gain antennas in the first place. The answer is simple: A radio wave from a high-gain antenna will attenuate at essentially the same rate as that from an isotropic radiator, but it will have a higher power to start with.

LOOKING UP

Now here is something you've probably never thought about, which clearly demonstrates how inaccurate some of our common conceptions about simple physical principles can be. Most astronomical objects can be considered as point sources (or reflectors) of light. Let's look at the Sun (using proper safety procedures, of course) to demonstrate a fascinating phenomenon. To illustrate how "rays" are emitted from the Sun, we're apt to draw a circle and a bunch of arrows pointing out *normal* to the circle (perpendicular to the surface of the Sun). If this were an accurate representation of a point source of radiation, to us Earthlings the Sun should appear as a minuscule pinpoint of light out in space. Only the specific ray that happens to be aiming right at Earth will be visible. So, why do we see our closest star as the familiar "Sun-size" disk?

Figure 8.1 shows our conventional concept of light rays (or radio rays) emanating from the Sun or other distant celestial object. (Again, we need to emphasize that the ray is merely a conceptual diagram). However, the Sun does not emit rays radially from its "geographic center." Rather it spews untold trillions of nearly isotropic radiators in the form of atoms in the *photosphere*, the gaseous atmosphere surrounding the Sun. Each "particle" of the photosphere radiates in an approximately isotropic manner, which means a certain number of rays

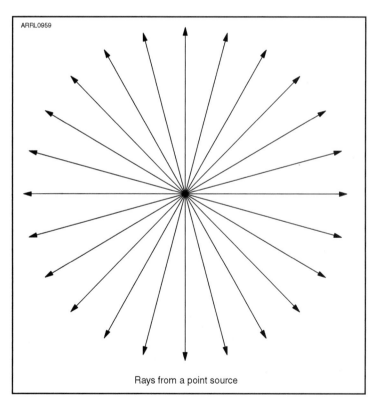

Rays from a point source

Figure 8.1 — This illustrates the common misconception of solar radiation as "point source." Unfortunately, this inaccuracy is often reinforced in a lot of literature on electromagnetic radiation.

Free Space Radio   8-3

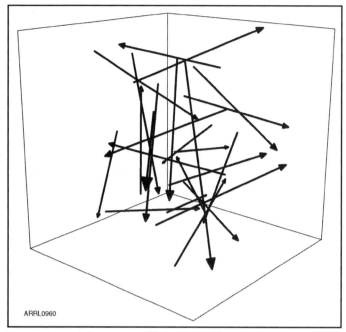

Figure 8.2 — A far more realistic view of how "rays" emanate from the Sun. The Sun's photosphere consists of countless trillions of such point sources, each emitting radiation in random directions and phases.

are going to be aimed in our direction at any given time (see **Figure 8.2**). If you have enough random radiators, statistically, a certain number of rays are going to be collimated, even though the vast majority of them will not be. So, the bulk of the radiation from the Sun appears as a very nearly Sun-sized object, from our point of view. Now, it probably goes without saying that it's a *very* good thing that the Sun is not a *coherent radiator* — meaning, the phase relationship between any two points in the radiation field has a constant difference, or is exactly the same in either the spatial or the temporal mode. If all that coherent radiation was aimed *at* us, laser-like, life would not be pleasant under the Sun's "glow."

Let's reflect on (or rather, off) an object closer to home. No doubt many of our readers are Amateur Radio moonbounce enthusiasts. Any practitioner of this art knows that the roughness of the Moon causes a phenomenon known as *libation fading* as various *non-coherent reflections* from the Moon's surface add and subtract in phase as the Moon wobbles slightly in its orbit. However, most avid moonbouncers don't realize that if the Moon were a polished sphere, moonbounce would be utterly impossible. It is the very roughness that makes the Moon as "big" as it is, from a radio viewpoint. If it were a polished sphere, the chances of a reflection coming right back at you would be about the same as a shooter marble or a billiard ball coming right back to you after hitting its target; the only time two polished spheres return to their original places after a collision is if they hit precisely dead center.

ON RADAR

If you were given the task of creating the perfect stealth aircraft, one that is invisible to radar, or nearly so, what would it look like? The answer to this was clearly demonstrated a few years ago by some visiting scientists from EISCAT at the University of Alaska's Geophysical Institute. These gentlemen suspended a one meter polished aluminum sphere from the ceiling of the auditorium. They then turned out the lights and passed around a laser pointer, asking people in the audience to try to locate the sphere with the laser pointer.

Even after knowing about where the sphere was, it was nearly impossible to locate it with the laser beam. Occasionally a very fleeting glimpse of red light could be seen after reflection from the sphere, and usually not by the person holding the laser pointer.

One might reasonably ask why the sphere was so easy to see with the room lights on. This is because the sphere is illuminated by countless scattered light sources, not a single ray.

This exercise clearly demonstrated the concept of *radar cross section*, which is a measure of an object's "size" when illuminated from a point source (such as a laser or radar). It also showed that the ultimate stealth aircraft would be a polished sphere, which, obviously, would be a bit difficult to fly, at least with known propulsion technology. (Evidently, only space aliens have figured out how to do this. Next to a polished sphere, the "classic" flying saucer shape would actually be a very good choice for a stealth aircraft. As long as there were no concave or flat surfaces, it would have a nearly non-existent radar cross section.)

We need to be familiar with the concept of radar (or just plain radio) cross section for any number of scientific studies. Some of our objects of interest are hard (like the Moon) and some of them are soft and squishy, like the ionosphere. The standard catalogs of cross sections generally assume highly reflective materials. On the other hand, the ionosphere is not only reflective, but also absorptive, which makes the cross section calculations much more difficult. A small, highly reflective object may have the same cross section as a large "fuzzy" object.

Although radio amateurs aren't allowed to operate radar *per se,* there are many radar-like methods we will use in radio science. One of the most important measurements we can make of radio signals is *time of flight,* the time it takes an object, particle or acoustic, electromagnetic or other wave to travel a distance through a medium. This requires some sort of *pulsed* or *coded* signal. The most common time-of-flight measurement used by radio amateurs is the *ionogram*, which we will explore in great detail in subsequent chapters on the ionosphere.

IS IT CLOSE ENOUGH FOR YOU?

In order to do time-of-flight measurements of radio signals, we have to know how fast radio signals travel. Most radio amateurs know that radio travels at 300,000,000 meters per second. It's a nice round figure, and it's *nearly* exact.

A lot of radar methodology and terminology has nautical roots, and some great rules of thumb have been developed by the US Navy's "Scope Dopes" (radar, sonar, or weapons fire control operators). One particularly useful one is "sea six," which is easy to remember because it's sounds like "seasick." This refers to the fact that radar pulses take 6 µs to travel 1 nautical mile. A nautical mile is 1852 meters, or about 6076 feet, which happens to be 1 minute (1/60 of 1°) of longitude around the equator. Sea six is not *quite* as accurate as 300,000,000 meters per second, but it's close enough for government work (at least Navy radar work). In radar, however, you have to account for the round trip, so it takes 12 µs for a radar pulse to get to an object 1 *nautical mile* distant and back again. To avoid confusion, remember what it says on the wide-angle sideview mirror of your car: "OBJECTS IN MIRROR ARE CLOSER THAN THEY APPEAR." Objects on a radar screen are also closer than they appear, by a factor of two.

For landlubbers, an equally useful figure is that radio travels at about 1 *statute mile* every 5 µs, or about 1 ft/ns, but this figure is substantially less accurate than sea six.

One may fairly question whether any rules of thumb have a place in precision scientific investigation. After all, we spent the entire first chapter preaching on the importance of accuracy. This is where we have to make the crucial distinction between *precision* and *accuracy*. Precision is the measure of how fine your measurements are, while accuracy addresses the absolute value of the quantity you're measuring. There's really no point in using a precision laboratory balance to measure bales of hay. There is no laboratory standard definition of what a "bale" is, so the precision of the measurement is wasted on such a fundamentally inaccurate unit, such as the bale.

When using time-of-flight measurements to measure the "height" of the ionosphere, we run across both *precision* and *accuracy* issues, which probably can't be (and probably don't need to be) fully resolved. In this case, rules of thumb are probably better than what we can measure. I'll explain this briefly here, but we'll go into more detail in our ionospheric chapters.

Because of limitations of available hardware and methodology, we're lucky to measure the height of the ionosphere to within approximately

5 miles, so we have a fundamental instrumentation precision problem to deal with. Moreover, regarding the ionosphere, height itself is not well defined. We have two basic ways of looking at it: the *virtual* height and the *reflection* height, as well as variations of both. Why? Because the ionosphere is not a mirror (there is no clear line of demarcation of where it begins or ends), so we also have a fundamental accuracy problem, simply for lack of a decent definition of the thing we're measuring.

But we need not despair. Many measurements in ionospheric science are *comparative*, meaning that all our samples have the same inaccuracies built in. In other words, many of the inaccuracies we deal with will cancel out, or become irrelevant, at least in terms of what we're actually looking for.

To offer a crude analogy of how this works, say you and I are riding along in a train, and I want to measure the time it takes for me to walk from my seat to your seat. Ignoring Einstein and relativity for the time being, you can see that it really doesn't matter how fast the train is moving to come up with that measurement.

PICKING UP SPEED

Speaking of speed, we can't spend much time studying electromagnetic propagation without running into the terms *group velocity* and *phase velocity*. Although there are aspects of these concepts that dig very profoundly and subtly into the very nature of existence, when it comes to practical measurements, we can make sense of this quite simply.

It's impossible to measure the time of flight of a continuous wave (CW) signal. Why? First of all, you can't even define it. If a radio wave (sine wave) has always existed, there is no time at which the "beginning" of the wave is in a different location from the "end" of the wave. However, you *can* measure the velocity of propagation of the wave itself, with some surprisingly simple apparatus, such as the Lecher wire. That is, assuming we know a few things about a continuous wave, the most important being that its period (the time it takes for a wavelength to pass a stationary point) is inversely proportional to its frequency (the number of wavelengths that pass a point in a set period of time). With a Lecher wire, you can measure the phase velocity of a wave; that is, the time it takes for the wave to reach the same amplitude on one cycle that it was on the previous cycle.

Measuring the time of flight of a pulse, however, requires that you *have* a pulse. A CW wave is not a pulse, which has a beginning and, presumably (but not necessarily), an end. We can measure the time of flight of the "turn on" time of our wave, since that turn on time has a definition. The speed at which the "beginning" of the wave travels is the group velocity. In free space, the group velocity and the phase velocity are pretty much the same, at least as far as we're concerned in the radio world. However, when we start moving through *anisotropic* (directionally dependent) media, such as the ionosphere, these can be two very different things. Furthermore, the velocity of propagation in a plasma (what the ionosphere is made of) is not the same as in a vacuum, so even the phase velocity is different. This is why we started this chapter on the free space properties of radio.

A Polar Exploration, Practically Speaking

Traditionally, radio amateurs haven't paid a great deal of attention to the polarization of radio signals. VHF and UHF enthusiasts are generally more aware of polarization than the typical HF denizens, primarily because such signals are generally line of sight, and their polarization is fairly well preserved over their entire paths. If two VHF stations happen to be cross-polarized, the results are obvious and dramatic. In theory, one can experience 100% signal loss by having a transmitting and receiving antenna cross-polarized (orthogonal); in practice, however, the maximum possible losses are likely to be on the order of 60 dB or so. In many cases, the ability to achieve extremely high levels of cross-polarization rejection is quite desirable. Perfect cross-polarization will allow the reuse of frequencies on a single frequency over the same path — although this technique is seldom applied in practice. However, careful cross-polarization will let you space communications channels much closer than you could with haphazard polarization, and is standard practice in satellite communications.

Circular-polarization (CPOL) is frequently used in amateur space communications, but generally not to its full advantage. Most hams who do use CPOL use it because it allows them to be a bit "sloppy" on the polarization alignment. A CPOL antenna will respond equally well to a signal of any *linear polarization*, though with a 3 dB power loss over a comparable, properly polarized linear antenna — a more than reasonable tradeoff for ease of operation. Not only are most amateur satellite signals of indefinite polarization, but at many popular satellite amateur frequencies, the polarization can be twisted by ionospheric effects, such as *Faraday rotation* (the rotating of the plane of polarization of a linearly polarized wave propagating through a magnetized dielectric medium).

On the HF frequencies and below, it's quite a different story. Most hams operate under the premise that HF signals lose their polarization sense after a "bounce or two," so they generally deem the polarization of incoming HF signals meaningless. From a practical communications standpoint, there is nothing particularly wrong with this approach; a century of successful shortwave communication using randomly polarized receiving (and sometimes transmitting) antennas is hard to argue with. However, for the purpose of radio science, the polarization of incoming signals tells us more about what's happening up there than just about any factor. The so-called random variations in polarization are precisely what the radio scientist is looking for in many cases. However, in order to make sense of the vagaries of the ionosphere, we need to know what our instruments are doing. For the radio scientist, the antenna is the primary instrument, so it's crucial to know what the characteristics of that instrument are. And among the most important characteristics are its polarization properties.

When the subject of HF antenna polarization is broached, many hams will argue that it's impossible to know the precise polarization of an HF antenna. They will argue that it is impossible to approach anything like free-space conditions at practical HF frequencies, since we have ground reflections, unknown direction of arrival, and other unknown factors. These arguments are valid, but certainly not insurmountable. In fact, we will demonstrate that all these factors are cancellable with proper construction and calibration methods.

ORTHOGONAL THINKING

Throughout this book, as well as throughout your scientific investigations, you'll see that the *complex number* system permeates just about any physical system we can think of. Complex numbers are useful abstract quantities that can be used in our calculations to obtain real-world solutions (see "When Numbers Get Complex" sidebar). For instance, we have numerous cases where we have two different parameters describing a system or circuit, such as resistance and reactance. These factors are plotted on *orthogonal* axes, meaning that it's possible to alter the value on one axis without affecting the value on the other. The circularly polarized antenna is no exception in this regard. In fact, there is an even greater symmetry in a CPOL antenna than there is in a complex electrical circuit, where the resistive and reactive components have rather different behaviors and significances.

You may find it helpful to think of any antenna as being circularly polarized, or, more precisely, *elliptically polarized*, but with varying

When Numbers Get Complex

There's something incredibly elegant and useful in expressing radio concepts in the form of complex numbers. Whether you're dealing with resonant circuits, circularly polarized antennas, or transmission line impedances, the complex number system is the most revealing (and sometimes only) way to adequately describe what's going on in a physical system.

For a closed circuit (lumped constant) system, the complex number is extremely intuitive. The *real* component tells you that something is going to get hot. The *imaginary* component tells you something is storing energy. Or, put another way, the real component tells us that something is doing real work, while the imaginary component tells us something is capable of doing some work, generally at some time in the near future.

Now, even if you *are* comfortable with complex numbers, there's one little fly in the ointment that threatens to make the topic a little more, well, complex. As it turns out, physicists and electrical engineers don't quite agree on what's real and what's imaginary. The issue at stake is that to the electronics person, real always designates *dissipated power,* or what is converted into heat, while imaginary designates either stored or *transferred* power. In the world of optics and more general physics, the real component designates power that is transmitted through a medium, while the power lost in the medium is the imaginary component. This is called the *extinction coefficient* by optics folks and others. There is a certain logic to the latter designations, in that transmitting through the media is generally what you're trying to accomplish in optics, while scattering and other "nonproductive" modes are deemed imaginary. For the physicist, real designates useful, while imaginary designates generally undesirable things (which are ultimately converted into heat).

This conflict shouldn't cause apoplexy, as long as you are aware of the two standards. We managed to survive a similar diverging of the minds about a century ago, when the raging debate was whether current flowed in the direction of electron movement, or in the direction of ion movement (or more recently, *hole* movement). Semiconductor symbols and such still point in the "wrong" direction.

---

levels of *ellipticity*. Even a perfect linearly polarized antenna may be described as an elliptically polarized antenna with an infinite *axial ratio*. The axial ratio of an ellipse is the ratio between the length and the "fatness" of the ellipse, and can be any value between 1, for a circle, and infinity for a straight line. Always remember that 0 is a number, too. The advantage of this approach is that we no longer have antennas that are special cases. They are all just different degrees of the same thing.

BACK TO OUR IDEALS

As in any other nook of radio science, it's important to describe the ideal before we tackle the unusual. So, in deference to our free-space

VHF/UHF folks, let's talk about how polarization affects antennas in free space.

Let's set up a resonant ½ λ transmitting dipole and a resonant ½ λ receiving dipole in free space, a few miles above the Earth, a dozen or so wavelengths from each other, so as to avoid any near-field effects (the only reason we need the Earth at all is for a reference point of "vertical" or "horizontal," which obviously have no meaning without some planetary body). In reality, it doesn't matter if the antennas are resonant or not, or even if they're remotely close to ½ λ, but this allows us to work with something standard and familiar. It also doesn't matter if the conductors are fat or thin, but for the sake of argument, we'll also assume very thin conductors — for all practical purposes an infinite length-to-diameter ratio. This allows us to know in which direction the conductors are oriented, with no ambiguity.

If the antennas are both vertical, they will have the maximum transfer of power between them. The direction of the conductor defines the direction of the *electric* field. As the electric field from an electromagnetic wave intercepts the receiving antenna, it causes electrons to accelerate up and down the conductor. Just to keep our axes straight, let's mount each antenna on a long insulated shaft, where each antenna can rotate only around the common shaft. This clarifies that we are altering only the relative polarization and not the *beam steering*. In other words, the antennas are not allowed to tilt toward or away from each other, but only spin around in their *mutually parallel planes*.

If we keep our transmitting antenna fixed, and rotate our receiving antenna around the shaft, we will find that the induced voltage drops off as a sine of the relative angle between them, eventually becoming 0 as it becomes horizontal. Or, more conventionally, we can say that the induced voltage is proportional to the *cosine* of the relative angle between them. Interestingly enough, this pure sinusoidal relationship is completely independent of antenna gain. We will find that a high-gain Yagi, or even a parabolic dish, exhibits the same sinusoidal polarization behavior — with one caveat: both antennas have to be *on axis*. If the antennas are not on axis, that is, if we allow the antennas to rotate in *non-parallel planes*, we will find that as we rotate the polarization of our receiving antenna, the pure sinusoidal relationship will not hold. (This is clearly visible with proper *NEC* antenna modeling, as well as through actual measurements). In addition, the polarization *sensitivity* will not be as great. If the antennas are not on axis, we will not achieve a perfect null by cross-polarization. In fact, we may find it a bit difficult to even define cross-polarization, since there are now two axes involved. So we can conclude that maximum *polarization sensitivity* will always occur on axis for any antenna.

NULL AND VOID

For on-axis antennas, it should be fairly obvious that the cross-polarization "null" point, that is, the orientation of zero power transfer, is much better defined than the maximum "correct" polarization peak. The "zero crossing" is an infinitesimally small, razor-sharp condition, whereas the maximum point of a smooth sine wave can be a bit hard to locate.

Most hams who do serious radio direction finding know that using the null method, such as with a small loop antenna, is a far more effective means of pinpointing a radio source than using even the highest gain practical Yagi. Nulls in radiation patterns can be *much* sharper than lobes. The real downside to using the null method is that the most effective null antennas have *two* nulls. The resolution of this ambiguity is usually a simple matter, however.

So it should be no surprise that the polarization null is more effective for finding the polarization of a signal than looking for the maximum.

SENSITIVITY TRAINING

Because the *off-axis* polarization sensitivity of an antenna is less than the on-axis sensitivity, we will find that this property is extremely useful for pinpointing the direction of arrival of a radio signal. In other words, we can "backwards engineer" the actual axis of the antenna, relative to the incoming signal, by rotating the antenna around the axis and seeing if we can get a perfect polarization null. Again, to keep our axes straight in our heads, let's mount a receiving dipole on a boom, precisely perpendicular to that boom. We want to find out the direction of arrival of a signal. Let's take a guess, and aim the boom where we *think* the signal is arriving from. If we rotate the dipole around the boom, we will find a perfect cross polarization null *if and only if* the boom is aimed precisely at the incoming signal. If we can't find a perfect null, we know the boom is off axis. We then need to re-aim the boom and try again, once more spinning the dipole around the boom, looking for a null. After a few more iterations, we should be able to nail the direction of arrival precisely.

Now, obviously, this can be a bit of a tedious exercise — and a bit impractical if you're trying to steer an 80 meter dipole. Fortunately, as we mentioned above, the polarization properties of an antenna are independent of its size — a 6 foot long loaded 4 MHz whip, if well balanced, can be used quite effectively. At HIPAS Observatory, we used such methods to do 3D triangulation of radio signals from regions of interest in the ionosphere.

Actually, we've killed two birds with one stone if we've done this right. Since a perfect null will occur only if we're cross-polarized *and* on-axis, if we find such a null we can know the direction of arrival *and* the polarization of the signal. Pretty slick.

Well, almost; there's only one small catch. This method works only if the incoming signal is linearly polarized. As we have already learned, this is seldom the case for ionospherically refracted signals, or for signals that have been *launched* (radiated) elliptically, either intentionally or otherwise. If an incoming signal is purely circularly polarized, we will see no change in signal strength whatsoever as we rotate our dipole around its axis. If there is some degree of ellipticity to the incoming signal, we may see some change in signal level as we spin our dipole, but we will never find a sharp null.

The only way to obtain a perfect null on a circularly polarized signal is by means of a circular polarized antenna of the *opposite sense*. Here *sense* means handedness, or the direction in which the electric field vector rotates. And, in fact, the circularly polarized dipole (or an array of them) is the standard receiving antenna in all ionospheric research, as well as ionospheric instruments such as the *ionosonde* and *digisonde* — devices used to measure the virtual height of ionospheric layers using a swept-frequency, pulsed radar technique.

There's good news, too. Have you ever attempted the tedious process of DFing a radio signal with a dipole (aim, spin, aim, spin, aim, spin…*ad nauseum*)? Interestingly enough, the circularly polarized antenna makes this process much easier. The circularly polarized receiving antenna already does the "spinning" for you, once for every cycle of the RF signal. All you need to do to pinpoint an incoming CPOL signal with a CPOL antenna is to *steer* the antenna for a null. And the incoming signal doesn't even have to be perfectly circular to take advantage of this. If there is any degree of ellipticity, you can still find a minimum by steering alone, even though it won't be a perfect null, except in the case of perfect *circularity*.

DON'T BE FAZED BY PHASING

Because of the general mechanical awkwardness of steering HF antennas, particularly CPOL ones that tend to occupy a lot more "sky," phased array methods are standard practice in ionospheric study. HAARP uses a large array of CPOL dipoles. Such methods are suitable for amateur experimentation as well, though on a much smaller scale, of course. With an array of just four crossed dipoles (or inverted Vs, which are even easier to hang) one can build a very effective instrument for all manner of HF propagation studies. Several hams have such versatile instruments on moderate-sized lots. In addition, these make fabulous near vertical incidence skywave (NVIS) arrays for emergency communications.

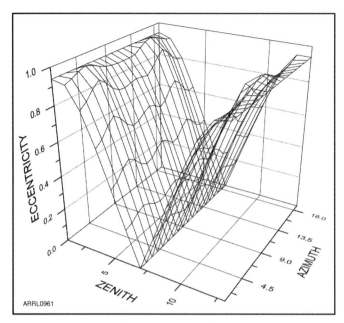

Figure 9.1 — This interesting plot shows the departure of circularity of a turnstile antenna. This is an important consideration as we use such antennas to perform more advanced radio science experiments.

One of the interesting characteristics of a planar array of CPOL antennas is that the composite circular polarization characteristic of a single antenna is maintained, regardless of the size. What this means for the average ham is that if you want to build a CPOL dipole, your elements don't have to cross each other in the middle, which can be a real benefit if you have a small or weirdly shaped lot. You can place a north-south dipole several wavelengths away from an east-west dipole, feed them with a 90° phasing line, and have a perfectly acceptable CPOL antenna.

Crossed dipoles are certainly not the only way of achieving circular polarization, but they are the most intuitive, and generally require the least amount of tweaking for good performance. As it turns out, nearly every antenna in a real-world environment has some degree of circularity in some direction — even the lowly dipole over real earth. This can easily be confirmed by *NEC* modeling, where the *axial ratio* is the indicator, where it's almost impossible to even model an antenna without some degree of ellipticity. Unfortunately, the actual volume of space over which these patterns are CPOL is generally very small, and not of great usefulness. To "go CPOL" one has to be fairly deliberate about it.

It helps to do a bit of *NEC* modeling to show how the *eccentricity* of a CPOL antenna varies as you move off axis, a plot of the eccentricity of an NVIS turnstile antenna is shown. Eccentricity turns out to be the reciprocal of axial ratio, which is a little easier to plot. As you can see in **Figure 9.1**, the best circularity is shown as 0 on the Z axis. The eccentricity doesn't change as you move around the azimuth, but it changes drastically as you move off Zenith. There's nothing like 3D graphing to clearly show this sort of thing. For a "real" example, see **Figure 9.2**.

Figure 9.2 — Roger Weggel, WL7LU, and Clayton Cranor, KL3AB, teachers in the Aero Sciences department at the University of Alaska Fairbanks Technical and Community College, examine the author's active, circularly polarized, broadband receiving antenna. [Eric Nichols, KL7AJ, photo]

NON-NULLING METHODS

While antenna nulling methods are very effective for pinpointing direction and polarization, the main problem is that they're *nulls*. They work best when they work worst. How does one make a signal disappear without making it disappear? For circular polarization, you will want some method of *sense switching*. This is easy to implement by just changing the polarity of one of the dipoles in a crossed dipole array. We showed how simple it is to use a "wrong-handed" CPOL antenna to locate the direction of arrival by null methods. If we use matched polarization, we will obtain maximum signal strength for on-axis signals. In addition, if the signal is *known* to be truly circular, or nearly so, we can determine the "aim" by looking for the maximum circularity of the received signal. As we steer off axis, a CPOL signal will become more and more elliptical.

On the other hand, if we don't know the actual circularity of the incoming signal, we can still use maximum signal strength to properly aim the antenna, and then we can use the antenna to determine the actual ellipticity of the incoming signal. This is the most useful method for ionospheric science. As you can see, being able to switch *senses* can do a lot to resolve various ambiguities that might show up.

When a radio signal is arriving from directly overhead, or nearly so, it's fairly simple to obtain quite accurate vertical angle of arrival information, especially with an array of NVIS antennas. This may be a bit surprising to most hams who pay little attention to the vertical angle of arrival, but are always conscious about azimuth. You certainly see a lot more azimuth rotators than elevation rotators on HF. Although transmitting *launch angle* seems to have gotten a lot more attention in recent years, there's still little being done in the way of *angle of arrival*. Based on the conventional *geometric optics* model, it is generally assumed that launch angle and angle of arrival are about the same — at least in the long haul. As we begin to fully recognize the non-reciprocal behavior of the ionosphere, however, we realize this is not the case. Nearly all ionospheric propagation consists of either a well-defined O mode wave or a well-defined X mode wave (clockwise CPOL and counterclockwise CPOL in the northern magnetic hemisphere, respectively). The X and O mode signals diverge not only in azimuth but also in vertical angle and reflection height. In addition, the ionosphere is seldom horizontal or uniform, but rather "bumpy" — sometimes to a surprising degree. A signal that starts out as a grazing path in one part of the world may arrive somewhere else as an NVIS signal.

FULL CIRCLE

What is the real importance of circular polarization in radio science — at least *ionospheric* radio science? As we've seen, ionospheric science is done mainly by inference; that is, we know what it is by what it does, sort of. One of the most crucial ingredients in probing the ionosphere "by proxy" is knowing which of the two characteristic waves we're dealing with. *Very* different physical and chemical processes are involved in the X mode wave and the O mode wave. Again, we encounter a difference in priorities between the typical radio amateur and the Amateur Radio scientist. The former generally wants to use whatever works to get from point A to point B, without a great deal of thought as to the intermediate processes. The latter wants to know what's happening that *allows* things to get from Point A to Point B, and sometimes this process is of far greater interest than any "intelligence" that might be conveyed from Point A to Point B.

10

The Big Picture — Plasma as Pachyderm

We've all heard the parable about the four blind men describing an elephant. One blind fellow grasps the elephant's leg and declares, "The elephant is like a tree trunk." The second blind man grabs the elephant's ear and says, "No, the elephant is like a leaf." The third blind man grabs the elephant's trunk and says, "I think the elephant is like a fire hose." The fourth blind man finds the elephant's tail and says, "I think the elephant is like a rope." All of these descriptions of the elephant are accurate, but with only those four descriptions, a fifth party would be hard pressed to infer from that what "whole elephant" looks like. The ionosphere is sort of like a big elephant in this respect.

Scientists at the EISCAT facility at Tromsø, Norway, recently calculated the mass of the entire ionosphere, and it came out to just about 1 metric ton. Think about that for a moment. The entire region of the Earth's atmosphere responsible for all the radio propagation we take for granted could be compressed and scooted into the corner of a garage on a one-man pallet truck. It's fascinating, but how can we know that, or rather, how do we infer that?

Our knowledge of the ionosphere is primarily based on how it responds to electromagnetic waves, which is how we make our inferences. Now, for the typical radio amateur, the only thing that matters about the ionosphere is how it responds to radio waves, or more practically, how radio waves respond to it. For the most part, we hams have ignored the chemistry and physics of the region, and as long as the ionosphere does what it's supposed to do — reflect radio waves — we are happy. But such an approach is not apt to lead us to a real understanding of what's up there. And as Amateur Radio investigators, we have some very good reasons to care about what goes on above our heads.

PROBING THE PLASMA

The universe is a plasma, at least the part we can actually see. The Coalition of Plasma Science (**www.plasmacoalition.org**) offers this succinct explanation the term:

> Plasma is often called the "Fourth State of Matter," the other three being solid, liquid and gas. A plasma is a distinct state of matter containing a significant number of electrically charged particles, a number sufficient to affect its electrical properties and behavior. In addition to being important in many aspects of our daily lives, plasmas are estimated to constitute more than 99 percent of the visible universe.

As difficult as it is to study the ionosphere by inference, at least it's a whole lot easier to study than the rest of the universe, since it's right in our back yard, astronomically speaking.

One of the things we've discovered is that the ionosphere is a *contained plasma*, well, sort of. Among the quandaries of nuclear fusion research is how to keep a plasma in a stable, confined area. If we can fully understand how the Earth's ionosphere can be as stable as it is without being "in a can" of some sort, we can perhaps make some real progress toward the holy grail of nuclear science: sustained nuclear fusion. More immediately, understanding the ionosphere tells us something about our environment and ecosystems.

The ionosphere sits on top of the "normal" atmosphere, and as such, responds to the atmosphere to some degree; there are even weather-like phenomena in the ionosphere itself. We may learn something about global warming by looking closely at the ionosphere. The lower ionosphere, especially near the D layer, seems to be especially sensitive to the global environment. At HIPAS Observatory, we did a protracted experiment with Cornell University looking at a peculiar phenomenon known as Polar Mesospheric Summer Echoes (PSMEs), which cause nearly specular radio reflections from about 90 km. PMSEs may be strong indicators of atmospheric pollution, particularly as it breaks down into ammonia. We do know PSMEs are very sensitive to high altitude temperature changes.

There are still lots of bizarre ionospheric phenomena that have yet to be explained. For nearly a hundred years, radio operators have reported encounters with *long-delayed echoes* (LDEs) and such (we'll be exploring these and some other interesting oddities in the next chapter). Yes, these unusual phenomena are real, and their existence does not require otherworldly explanations. In fact, they can be reproduced in plasma chambers on smaller scales. The real question is what naturally occurring conditions cause them. We also have reason to believe that *parametric amplification* can occur, where radio signals are actually increased in intensity over their journey through the ionosphere, under just the right conditions. We now know that lightning-related phenomena extend into the very upper reaches of the ionosphere, as well. We have weird and wonderful things like jets and sprites that have been observed only in the last decade or so. And we suspect we'll make more related discoveries before too long.

APPROACHING REAL RADIO SCIENCE

In the early chapters of this book, we spent some time demonstrating how electricity interacts with matter, primarily from a chemical point of view. Now let's apply these principles to the concept of electromagnetic waves. Although some of the associated equations can be extremely daunting, the actual hardware for doing genuine radio science can be surprisingly simple, and more importantly, intuitive. Since we want to make radio science accessible to radio amateurs, we will keep the math to a minimum. If you can handle high school trigonometry and can think fairly well in three dimensions, none of the following concepts will be difficult at all.

When it comes to electricity, the electron is the prime mover (or "movee"). Not much happens in radio either, without the movement of an electron. It's actually the *acceleration* of a charge that's of interest to us, radio-wise, not the mere movement of the charge. For most of us radio amateurs, it's the acceleration of the electron that's of greater importance, although as we will learn later, there are some interesting interactions with ions as well. We don't usually see these ion interactions directly, however, because the mass of these ions is thousands of times heavier than the electron, so their acceleration for a given EMF is going to be relatively insignificant. The very lightest ion we'll ever encounter is the proton, which is 2000 times as heavy as the electron, and even these are of little interest to us. There aren't a lot of free protons floating around in the ionosphere. Most ions of interest in the ionosphere are oxygen and nitrogen ions (in various flavors), which are very much heavier than the proton, and thus much less "wiggle-able."

In most Amateur Radio experience, what we see as ionospheric propagation is the result of the acceleration of a vast number of *free* electrons. A radio wave encountering free electrons accelerates the electrons in the direction of the electric field of the radio wave. These electrons, as a result of their bulk acceleration, generate radio waves. This marvelous, reciprocal interaction between the wave and the "electron mob" results in what we call reflection. Reflection is really a bad term, since the process is really one of absorption and re-radiation, but it's sufficiently useful to describe what we observe on the ground. Therefore, for all its shortcomings, nearly all that we know about the ionosphere can be interpreted in terms of "reflections." But before we delve into more of this mob behavior, let's stop and talk a little about individual electrons.

To the astute and experienced radio practitioner, reflecting a radio wave off a single electron seems rather implausible. However, there is a scientific instrument that does just that; it's called an incoherent scatter radar (ISR). (There was one in the Fairbanks area several years ago, which had a monster 60 foot steerable dish and consumed great gobs of electrical power from our paltry power grid. It has since been decommissioned, but while it was operational, it generated a great deal of interesting science.) The real charm of an ISR isn't so much that it can see reflections from single electrons, but rather that it can see reflections or "scattering" from *bound* electrons; that is, electrons that are still in orbit around their host atoms. In other words, it can obtain reflections from a non-ionized region of sky.

Envision an electron in orbit around a nucleus. It stands to reason that at some point the electron is going to be moving away from you, and at other times it will be moving toward you. If you bounce a radio signal off one of these electrons, you can determine whether the electron is moving toward you in its orbit or away from you, and how fast it's doing it — that's our familiar Doppler shift. By measuring the speed of orbit of the electron around its core we can now determine the temperature. Higher temperature particles wobble faster than lower temperature ones. Since we're talking about radar, after all, we can also determine the distance to the region of interest. By plotting the temperature versus distance of our ISR return signals, we can develop a temperature profile of the upper atmosphere.

Naturally, no matter how clever we are, we aren't going to be able to truly look at individual electrons at a great distance, so we have to do a lot of statistical gyrations to come up with an average electron velocity. This is where the term *incoherent* comes from. We're still observing the return signals from individual electrons, but they aren't *coherent* as they would be if they were free electrons, all moving in lockstep with an impinging radio wave.

Not only can the ISR determine the temperature of the electrons at various heights, but by looking at the frequency spectrum and polarization of the return signals, scientists can determine the actual chemical makeup of the sky at various levels. Electrons have different shaped orbits around different elemental atoms, so their acceleration profiles will be different depending on the shape of the atom.

The ISR has helped confirm some of our educated guesses about the nature of the ionosphere, which we've derived from more conventional methods. It's one more tool we use to get another description of the elephant, to bring us closer to the whole picture.

MOB BEHAVIOR

Most of what we will be dealing with, at least as far as ionospheric studies are concerned, will involve the collective behavior of free electrons. It is to this end that most of our efforts and instrumentation will be directed. We'll offer a cursory review of some of the basics of ionospheric reflection, or more properly refraction, and then cover it in much more depth later.

The main ingredient of a usable radio ionosphere is a large quantity of free electrons. Where do they come from? They come from the atoms, which are in the upper atmosphere, primarily the aforementioned oxygen and nitrogen atoms. Radiation from the Sun, primarily ultraviolet, knocks electrons out of the outer orbits of these atoms. What's left behind are ions, positively charged atoms of the same type.

Now, we all know that electrons are mutually repulsive, and without any other influence, vast mobs of electrons are going to repel each other and drift off into oblivion. Electrons in oblivion aren't going to help us much as far as radio propagation is concerned. We need electrons in a semi-organized condition to act as a good reflecting surface. This is where the leftover ions come in handy. The ions, which are pretty much the same mass as the atoms from which they were created, are going to pretty much stay put. They are held in place by the same forces that keep the air itself in place, namely gravity. Because they are now positively charged, they have an attractive pull on the electrons in their vicinity. The net result is that the electrons arrange themselves in what amounts to a layer. It's wrong to conclude that this layer has nothing but electrons in it; rather, it is a region that has statistically *more* free electrons than other regions of the atmosphere. A function called the *electron density profile* shows the relative concentration of free electrons versus altitude, and it is clearly illustrated on any Digisonde ionogram. The peak of the electron density profile is generally around 250 km in height, during the daytime. This changes depending on where we are in the 11-year solar cycle, as well as on a few other factors, but it's a good round figure.

The maximum electron density has a few interesting, special properties. It occurs at the same height as the *critical height*; that is, the height of reflection at the *critical frequency*. Critical frequency, in turn, is the frequency at which all reflection ceases and vertically traveling radio signals pass clear through to outer space. The critical frequency, and the maximum electron density, follows the elevation of the Sun in a fairly sinusoidal fashion. If you look at a plot of the critical frequency, it follows an almost clockwork diurnal variation — *if* there are no other disturbing influences, such as magnetic storms. (See **www.haarp.alaska.edu/cgi-bin/digisonde/scaled.cgi**.) However, that's a relatively big if, and it's rare to see a sinusoidal density plot beyond a few days. Many unknown influences can create huge amounts of absorption in the ionosphere, and these events can be surprisingly sudden. These sudden absorptive events subtly suggest high-energy particle collisions with the Earth, and radio amateurs are uniquely equipped to measure and detect them.

YOUR RADIO SCIENCE TOOLBOX

As a typical radio amateur, you won't have an ISR in your radio science toolbox, but let's talk about what you probably already do have, and how we can optimize it for doing some serious radio research, above and beyond just communications.

The overwhelming majority of the methods you will be using to do radio science will involve receiving radio signals of wide-ranging frequencies and power levels. You don't need the most sensitive radio receiver in the world, nor do you need the biggest antennas. What you will need, however, is accurate information on the characteristics of your receiving apparatus. This, in itself, represents a major change in the way hams go about their business. When doing scientifically valid ionospheric studies especially, you need to know the following quite accurately, at a bare minimum:

• *The direction of arrival of radio signals, both in azimuth and in elevation.* The latter can present some interesting challenges at HF radio frequencies, to say the least. We will discuss several methods of doing this with a fairly high degree of confidence.

• *The polarization of incoming radio signals.* Many hams are surprised to learn that nearly all ionospheric radio signals are circularly polarized, and that there are two characteristic waves for any HF signal through the ionosphere: the X-mode signal and the O-mode signal. These two signals have opposite "handedness" or "chirality" (not superimposable mirror images). We need to positively identify which mode we're dealing with at any time.

(The use of circular polarization can significantly improve day-to-day HF communications effectiveness as well.)

• *Where you are*. We'll be looking at projects that involve detailed sky-mapping of the contours of the ionosphere over large areas of the Earth. It helps to know where you are to do this accurately; with universally available GPS, this is no longer as difficult to determine as it once was.

• *The phase of incoming radio signals*. For skymapping, you will need to employ absolute phase measurement, a moderately difficult task. For other experiments you might engage in, you may only need to measure instantaneous, relative phase shifts.

Most of the radio science experiments we'll be describing will use pre-existing radio transmitters. There are countless shortwave broadcast stations operating all over the world, 24 hours a day, with well-known and consistent transmitting characteristics. These are of inestimable value in performing really good radio science. Your transmitter of choice will often be your friendly local AM broadcast station, for a number of reasons.

Most radio amateurs are familiar with radio beacons. These are transmitters used for identifying band openings, usually in the higher frequency HF and VHF bands, where propagation paths are relatively rare. The channel probe is simply a beacon station with a dedicated, matching receiving station, capable of continuously logging propagation paths between pairs of points on the Earth. Channel probes are particularly useful for "catching" fleeting, intermittent propagation events. We could certainly stand to have a few channel probes on the amateur bands, as they require no special licensing.

The use of actual radar methods on the amateur bands, on the other hand, is somewhat limited. Pulse methods are allowed only on some UHF and microwave bands. However, FM or phase-coded CW radar is technically allowed wherever phase modulation is allowed, though these require a bit more sophistication to operate than a conventional "ping and sing" radar. (Some impressive recent technical advancements may mean that FCC Part 15 unlicensed ionosondes could become available in the near future.) Scientific radar includes everything from the ISR described earlier to the ionosonde, the standard scientific instrument for HF propagation testing. There are enough commercial ionosondes around, with readily accessible data, so we don't have to worry too much about building our own. Every ham should know how to read the ionograms they produce.

WWV, the NIST standard time and frequency station (actually several stations), serves quite nicely as an HF radar for some purposes as it transmits "ticks" at precise intervals, which work very well for doing time-of-flight measurements of ionospheric paths.

In Chapter 19, we'll describe a very compact, circularly polarized HF antenna that can be used very effectively for skymapping — as well as actual amateur communications.

The Complex Ionosphere — Weird and Wonderful Non-Linear Phenomena

In Chapter 10 we introduced some of the most important concepts we'll cover in radio science. Now let's begin turning the most central of those around a bit further to look at them from a few different angles.

As we've seen, ionospheric studies can be basically divided into two broad categories: linear and non-linear behavior. Just because something is odd or unfamiliar doesn't necessarily mean that it's a non-linear behavior. A lot of radio behavior may seem odd to the uninitiated, but after a bit of experience it becomes more logical and predictable; it just may not be totally uniform. For example, most experienced hams know that long-distance radio propagation is highly dependent on frequency. This is why we have different radio bands. But once we learn the rules, we realize that a lot of this is pretty much like clockwork. We get Result B by performing action A. This is the best definition of linear behavior, from a qualitative viewpoint. There is a direct correlation between Action A and Result B.

Furthermore, from a quantitative viewpoint, linearity suggests a certain proportionality between cause and effect. If doing Action A results in Response B, then doing more of A results in more B. Most of what we do is based on supreme confidence in the above (whether founded or not). The world is much more frightening when Action A results in response B, C, or D at random, or when more of A sometimes results in more of B and sometimes less of B. Welcome to the world of plasma physics.

Believe it or not, the neighborhood is not entirely strange, and we actually know quite a lot. Most of the Amateur Radio literature on ionospheric propagation is based on decades of experience and observations. Considering that we have been aware of the ionosphere for well less than a century, we are far better at predicting its behavior than meteorologists

are in predicting the weather. To get the discussion rolling, let's look at some undisputed facts:
- The ionosphere is full of ions
- The ionosphere is composed of discrete layers of ions
- The ionosphere is full of free electrons
- The ionosphere is composed of discrete layers of free electrons
- The free electrons and ions in the ionosphere are very closely matched in altitude
- The ionosphere has a different refractive index than a vacuum
- The changes in refractive index of the ionosphere closely track the electron density
- Compared to the ground level atmosphere, the ionosphere is extremely thin
- Free electrons are primarily released by ultraviolet radiation from the Sun
- Ultraviolet radiation from the Sun is directly related to sunspot activity

Crucially, the above facts have a very strong correlation, and not only do they correlate very strongly with each other, but also with observed radio behavior at "ground level." We have lots of Actions A high above us that fairly consistently result in Responses B here below.

Without the Sun, we wouldn't have a lot of radio. As we saw in our discussion about free space radio, the Earth intercepts a minuscule portion of the total radiation from the Sun (the "dc to daylight" we all know). The most obvious radiation is light and heat; the most significant non-obvious component is ultraviolet — that is, until we see some of its obvious results. Ultraviolet light knocks electrons off the atoms that make the atmosphere, primarily nitrogen and oxygen. Although other elements, such as noble gases, are present (and even much more easily ionized), they are extremely minor players in our topic of interest, because of their relative quantity. So we have primarily oxygen ions and nitrogen ions as a result of the electron loss. There are several varieties of nitrogen ions, which accounts for some of the subtler layering of the ionosphere. These different nitrogen varieties have a more profound effect on the optical nature of the ionosphere, however. Nitrogen has more possible atomic resonances than any other element in the ionosphere, and is capable of creating numerous modes of light generation, which accounts for many of the different colors of the aurora.

The freed electrons never travel far from the atoms from which they are separated. This has been confirmed by actual samples of the ionosphere taken by scientific rockets. The local ion density at any altitude nearly exactly matches the electron density at the same altitude. Because of this, we can get a very good idea of the ion density from ground-based radio measurements, even though radio does not interact with the ions directly to any degree (with some exceptions, as we'll see).

Electrons are extremely light and agile, compared to their associated ions. However, the ions are extremely important because they act as anchors for the free electrons. Without the anchoring effect of the ions, the electrons, being mutually repulsive, would wander off into oblivion. It is the ions that compel the electrons to align themselves in layers.

Right off the bat, we see that we have two very strong correlations established: the correlation between the ultraviolet radiation and the free electrons, *and* the correlation between the free electrons and the ions, both in quantity and altitude.

We need to make it clear that there is no region of the ionosphere where we have nothing but free electrons, or nothing but ions. At any altitude, we have a mixture of ions, electrons, and neutrals. The density profile simply shows us the relative proportion of free electrons we have at any altitude. In addition, we never have a fully ionized region; that is, an area where we have nothing but ions and electrons, but no neutrals. This fact becomes more important as we explore some of the odder behaviors of the ionosphere.

REFLECT ON THIS

Although long-distance radio propagation may seem to be the result of a simple reflection, we almost need to remove the term *reflection* from our vocabulary. *Refraction* is the better term, because it describes the *frequency dependent* character of the ionosphere more accurately.

The *refractive index*, a measurement of the bending of the electromagnetic wave, of a plasma increases with increasing electron density. In other words, the velocity of propagation is less going through a plasma than it is in free space. Now, if we could somehow generate a region of the ionosphere that was an absolutely perfect conductor —say, a massive sheet of electrons immediately above a region of free space (or neutral gas, which is about the same as free space) — we *could* have a perfectly specular, frequency-independent reflection. We sometimes approach this sort of reflection during E_s conditions, which are evidently caused by massive electron clouds far outnumbering the neutrals, at least in a small region of sky (we'll set this anomalous condition aside for a while, but we will revisit it).

Now we need to note that the electron density does *not* make an abrupt change from zero electrons to a whole lot of electrons at "normal" ionospheric altitudes. We find that there is a gradient of electron density with respect to altitude, which is known as the electron density profile and can be seen as a black bell-curve trace displayed on typical ionograms (**Figure 11.1**). This curve shows the relative density of free electrons versus height in kilometers. It is quite typical in shape for most of the world, but the peak of the curve can vary drastically from day to day and region to region.

If we send a vertical incident radio wave into this region from the surface of the Earth, the wave will gradually slow down, reverse, and then accelerate downward as it encounters decreasing electron density (and refractive index). There is a *real* height of reflection, which is the point at which the electric and magnetic fields reverse, and the wavefront reverses direction. There is also a *virtual* height, which is the height at which the reflection seems to occur, because the wave has slowed down through its journey. This virtual height is always higher than the actual reflection height. For a vertical incident wave, the only difference we'd see between these two heights is a slightly different time of flight. However, we can never actually see the real reflection height, because our time-of-flight apparatus (a radar, actually) only looks at the time delay of the radio burst we send up (we've returned to that group velocity thing).

All this changes when we send an *oblique angle* (not a right angle or any multiple of a right angle) radio signal into the sky, which is what all practical Amateur Radio communications are. The skip distance is dependent on the virtual height at the apex of the "reflection" (again the term used advisedly). However, there's one more complexity we need to consider. Not only does the refractive index of the ionosphere change with

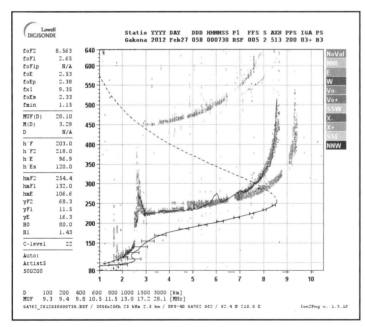

Figure 11.1 — A typical Digisonde ionogram taken during relatively stable conditions in interior Alaska. Ionograms in your neighborhood will likely look quite different, but the crucial information is all there. The most important figures are the critical frequency (first entry on the left-hand column) and the MUF figures, running just below the X axis.

height, because of the electron density gradient, but the refractive index is also dependent on frequency. Higher frequencies penetrate the plasma farther before reversing direction. So now we have a real reflection height that is dependent on frequency, as well as a virtual height (the one that really matters from a practical standpoint), which varies with frequency. This means that for any launch angle, the skip distance will be greater for a greater frequency. That is, up until we reach the *maximum usable frequency* (MUF), which is the frequency at which the signal goes off into outer space.

Looking again at Figure 11.1, we notice some other interesting correlations. For any vertical incident signal, we have a *critical frequency*. This is the lower trace beginning on the left of the ionogram. (The upper trace is just a multiple reflection, an indication that absorption was very low at that time) We see that at about 8.6 MHz the trace goes straight up. This is the critical frequency. But notice the height at which this makes the sudden upward bend. This is known as the *critical height*, and it always occurs at the same height as the peak of the electron density profile.

One of the more intriguing things about the critical height is that we can never get any ionospheric information above that point. The only way you can see anything there is by means of a so-called *topside sounder*, which is a downward looking ionosonde riding on a satellite. If you look at the electron density profile again, you will notice that it is a dotted line above the peak. This is because this density is only *calculated*, not *measured*. The ionospheric plasma totally shields the "top side" from any radio waves entering from below.

Now, as complex and convoluted as all this may seem, it all still falls under the category of *linear* ionospheric phenomena. It is all normal and predictable. And, more importantly, for the purposes of the following discussion, this behavior is *independent of power levels*. A radio amateur operating a QRP (low-power) station will observe all the same ionospheric behavior as the ham running a Big Gun station. Critical frequency, critical height, MUF, skip distance, reflection height, and virtual height are *all* linear functions, independent of power levels. The power you get out of the ionosphere is directly proportional to the power you put into the ionosphere. As we said at the outset of this chapter, more A results in more B, with some interesting exceptions.

When it comes to bizarre Amateur Radio observations, probably nothing is as mysterious — and persistent — as reports of long-delayed echoes, or LDEs. Since the beginning of radio, practitioners of the fine art have regaled us with stories of hearing their own signals return to them seconds, or even tens of seconds later. These reports are too numerous to ignore. These reports are too numerous to ignore. Seasoned seafaring radio telegraphers have reported them, astronauts have heard them, respected radio amateurs have heard them, and *I* have heard them.

All "normal" terrestrial explanations for LDEs come far short of explaining the long time elements observed in LDEs. Speculation that these signals are somehow ducted around the Van Allen belts is a bit farfetched. Multiple trips around the ionosphere *might* explain some of the shorter time element LDEs, except for the fact that these signals can be surprisingly strong. Some of the reports would require a trip to the Moon and back to account for the time delay. So far, the only HF moonbounce ever achieved was with HAARP — using a 1 GW on 40 meters.

When faced with such a multiplicity of speculations, it's always best to return to what you *do* know about such matters. Look for the answers closest to home before looking for extraterrestrial causes. We know a lot about radio propagation. We know a lot about ionospheric electrons. In fact, we know a lot about ionospheric *ions*, but we tend to forget about them when we talk about radio.

We have known for a while that most of the non-linear phenomena we see in the ionosphere are due to the direct interaction of radio waves and ions. The reason we don't normally see such phenomena, at least at Amateur Radio power levels, is that we just don't have enough muscle to move ions around — at least not very fast. Herein may lie the answer. Radio waves interact with ions just as they do with electrons; it's just that we normally don't see the effect, since ions are *thousands* of times heavier than electrons.

Since the discovery of the Luxembourg Effect (see "The Luxembourg Effect Put to Practical Effect" sidebar), scientists have wondered if there was some absolute, well-defined point at which non-linear phenomena, such as frequency mixing, occurs. Or does it occur *to some degree* at any power level, but not enough to notice until you get to megawatts of power?

One of the last collaborative experiments HIPAS and HAARP did was precisely to answer this question, with a power ramping experiment. We set up some sensitive ELF detectors, positioned about halfway between HIPAS and HAARP and powered up both of our transmitters a few kHz (in the low VLF range) apart, with a nominal carrier frequency of 4.5 MHz, and looked for the difference frequency with our ELF detectors. HIPAS kept pedal to the metal while HAARP ramped its power down from about 100 MW ERP to nothing. As I recall, HAARP was down to less than 1 kW before we could no longer detect the VLF difference frequency. We could see no "thresholding" effect whatsoever — the ELF

The Luxembourg Effect, Proving the Exception

In the year 1933, when high-powered shortwave radio first came into existence, an odd phenomenon was discovered, which would later be called the *Luxembourg Effect*. Radio Luxembourg, at the time one of the most powerful stations in existence, was mysteriously being heard on frequencies far removed from its licensed frequency. This was not on a harmonic of the radio transmitter. It turned out to be a *mixing* product between Radio Luxembourg and a Swiss shortwave station. But where was the mixer? Where was the non-linear device that was creating the difference frequency between the two radio stations? Both stations were carefully inspected for defects, but none were found. It turned out that the ionosphere itself was the culprit!

Subsequent tests demonstrated that the ionosphere, if "pumped" hard enough, could exhibit non-linear behavior. To further investigate this phenomenon, beginning in the early 1960s, various high-powered "ionospheric heating" facilities were constructed, first in Russia, and then at various places around the world: Tromsø, Norway; Arecibo, Puerto Rico; Fairbanks, Alaska; Gakona, Alaska (HAARP); and others. These facilities remain at the forefront of plasma physics research. Investigations into the Luxembourg Effect have led to a large number of fields of investigation within the realm of plasma physics, ranging from nuclear fusion to astrophysics.

One of the more promising practical applications of the Luxembourg Effect is in the field of ELF radio generation. Submarine communication is limited to radio frequencies below around 200 Hz, since these are the only frequencies that will penetrate seawater. *Skin effect* (the tendency for high-frequency currents to flow on the surface of a conductor) limits the penetration of HF radio frequencies to a few inches.

Previous methods of submarine communication involved vast arrays of wire antennas and vast amounts of transmitted power. Project Sanguine in the Lake Michigan area was the first successful implementation of this, but it was abandoned as too expensive and environmentally disruptive and an alternative was actively sought.

HIPAS Observatory, near Fairbanks, and HAARP in Gakona, about 300 miles to the south, were the first facilities to experiment with using the non-linear mixing properties of the ionosphere to generate ELF using HF transmitters as the "pump" frequencies. By sending two multi-megawatt HF signals (relatively easy to generate, compared to ELF) into the ionosphere, separated by a couple of hundred hertz, investigators found that the ionosphere produced both the sums and difference frequencies of the two transmitters. The sum frequency was of secondary interest, but the difference frequency in the ELF range was of great interest. This became the main driving force behind building the much bigger and powerful HAARP facility, after investigators had performed the original "proof of principle" at HIPAS.

signature just seemed to drop proportionally with the pump power. Just as a confirmation, both facilities ramped power down in the same manner; again, the non-linear effects seemed to track the pump power quite smoothly.

Granted, our receiving instruments were many orders of magnitude more sensitive than the average ham's, but it was fairly conclusive that any ham could be a HAARP! One of the problems is that such mixing products at amateur power levels would be well below the noise floor, but the fact is they would still be there.

We've seen that it takes a lot of power to interact directly with ions, at least enough that you'd notice, but we haven't talked about *indirect* interaction with ions!

HEAVENLY REVERB SPRINGS

The phenomenon of *ion acoustic* waves is well known in the plasma physics field. Ion acoustics is very similar to normal acoustics, but a great deal faster. "Sound" waves are transmitted through plasmas by the bumping of like-charged ions against each other. However, unlike "normal" air molecules, ions can also respond to radio waves, because they have an electrical charge.

Now, recall how we described the ionosphere as being layers of ions in close proximity to layers of electrons. If a radio wave intersects a layer of electrons, these electrons will oscillate in unison, in concert with the impinging radio waves. But the "mob" of electrons is *also* loosely coupled to the layer of ions. As the electrons "slosh" back and forth, they tend to drag the ions back and forth with them, but at a much slower pace. These wobbling ions can now propagate an ion acoustic wave, which is much slower than a radio wave, but perhaps 10,000 times the speed of sound, or around 2000 miles per second. An ion acoustic wave could make a lap around the Earth in about 12 seconds.

But here's the really intriguing part: Because ion acoustic waves are propagated through charged ions, these moving ions can create radio waves *at any point* in their journey. Think of this as a giant, natural set of reverb springs. You have an electrical to mechanical transducer at one end (radio to ion interaction), a mechanical delay system (ion acoustic waves), and another transducer (ion acoustic waves back to electromagnetic waves).

This entire process may seem a bit complicated, and, frankly, *unlikely* at first, until you realize that this exact same process can be consistently reproduced in a plasma chamber, which brings us to the subject of our next chapter.

Ionospheric Science in a Can

Doing ionospheric science is a mixed bag of nuts. The good part is, the ionosphere is very inexpensive — when it exists. The bad part is, it can be a very temperamental beast. In fact, the ionosphere has so many variables that its study can be very difficult and uncontrolled at best. Sometimes the best solution is to *build* your own ionosphere.

At various junctures in this book, we've used the terms plasma and ionosphere pretty much interchangeably. This was no accident, but it probably merits some clarification. The ionosphere is a special type of plasma, in a special type of container. Unlike the plasma in your plasma screen TV, the only container for the ionospheric plasma is gravity. Plasma physicists, when comparing the ionosphere to their "normal" plasma lab methods, often refer to the ionosphere as a "plasma chamber without walls." The walls of a plasma chamber always alter the behavior of a plasma, and so laboratory physicists studying it often look enviously at those who work in facilities such as HAARP, EISCAT, or Arecibo with their unfettered access to the plasma surrounding the Earth. But, conversely, many ionospheric physicists relish the opportunity to get their hands on a plasma once in a while and choke it into submission.

Don't worry, we don't expect the average ham to be able to build a large plasma chamber in his or her garage, but a small one can be built with not much more than a vacuum pump and some plexiglass tubing! Since nearly every ham eventually *uses* a plasma in one form or another, it's well worth the time to learn some basic plasma physics principles. At the very least, the inquisitive ham should *know* what folks are doing in plasma labs, since this promises to be one of the biggest fields where hams might use their skills professionally.

We'll describe some of the more recent plasma chamber experiments that fairly conclusively show that phenomena such as LDEs and parametric amplification can (and probably do) exist "in the wild"; that is, naturally occurring in Earth's ionospheric plasma. Everything that follows can be understood by anyone who's been able to pass an Amateur Radio license exam, and all the associated hardware is so basic you could draw it with a fat purple crayon — that's my gauge of simplicity!

THE PRESSURE'S OFF

Let's look at how a small, rudimentary general-purpose plasma chamber can be put together. While it's not sophisticated like you'd find in high-end facility (**Figure 12.1**), it will easily illustrate basic plasma physics and pretty well simulate most ionospheric conditions with a conventional vacuum system.

Figure 12.1 — This inductively coupled plasma (ICP) chamber is a little advanced for the typical garage constructor, but it clearly shows the components common to all plasma chambers. Notice the induction coil surrounding the glass vacuum chamber. Radio frequency energy is fed into the coil, which then heats the plasma. [Photo Courtesy of NASA]

At its core is a thick-walled Plexiglas cylinder, around two inches in diameter and about eight inches long. At either end of the cylinder is an aluminum end cap, with an O-ring coated with vacuum grease (which does not "outgas" when subjected to a high vacuum) sealing it against the plexiglass tubing. One of these end caps has a vacuum hose fitting on its edge; both aluminum end caps also have what's known as a *probe port*. This is a vacuum-sealed, electrically insulated fitting in the center of the end cap, through which one can poke an electrode or other sensing device. Larger and more elaborate plasma chambers may have servo-driven probes, capable of extremely precise positioning. The mechanism of choice for this is a *linear actuator*, which is a stepper motor running a jack-screw mechanism. However, for our immediate purposes, our fingers will serve as the linear actuator. A vacuum hose running from the end cap hose is fitted to a standard *positive displacement* vacuum pump, which is not much more than a piston compressor operating in reverse (*extremely* high vacuum would require *diffusion pumps* or other similarly exotic devices).

We now insert a couple of sharpened spikes (ideally made of tungsten, though any reasonably hard material like nichrome or stainless steel will do) through the insulated probe ports. In the lab, we start with ⅛ inch rods of the material in question and grind them on a grinder to a lethal looking point, positioning the tips of the probes an inch apart or so (one beautiful thing about plasma physics is that you can do a lot of great science with an "or so" level of precision). Next we connect a high voltage dc power supply to the spikes. High voltage in the plasma lab can mean anything from a few hundred volts to millions, but we'll be working at the low end for the sake of this discussion and a couple of hundred volts between the spikes is a good starting point. With atmospheric pressure in the chamber, nothing will happen.

It's now time to turn out the lights and start the vacuum pump. At first, the pump will make a "pocketa-pocketa" sound until the pressure drops to a low value, at which time it will become eerily quiet. Looking at the sharp points of the spikes, you would see a faint purple glow around the area of either point. If you're really observant, you would notice that one of the "glows" looks just slightly redder than the other. This will be the positive electrode. There might be a slightly perceptible difference in the *shape* of the glow as well.

What is happening? Well, the positively charged spike is "sucking" extra electrons out of the rarefied air in the vicinity, leaving behind positive ions. Conversely, near the negative spike we have the production of free electrons, which are boiled off the "cold cathode." We may have some negative ions in the region as well, but these generally require higher voltages to produce. Curiously, this behavior occurs even if there is no current flow *between* the electrodes; in other words, you don't need an *arc* to have a glow discharge.

Now, as long as the two electrodes are spaced adequately from one another, we have two individual plasmas: a decidedly positive one and a possibly negative one. But there's also more happening than meets the eye. We know from quantum theory that light in the form of photons is emitted from atoms when electrons in an excited state return to a lower energy state. It's not the ionization itself that creates the visible light. Additionally, an ion can return to a less excited state and still be an ion. Ionization and excitation are two different things: Ionization is removing an electron from orbit around its host atom, while excitation is kicking lower level electrons into a higher orbital state. You generally have both processes occurring, however, when ionization is happening. It's hard to ionize an atom without exciting it, too. Just be aware that visible light is an *indicator* of what's happening in the ionization process, but it's not the thing itself. You also need to be aware, especially as with more complex atoms, that most atoms have *more* excitation states than ionization states. Nitrogen, for example, has *many* excitation states, each of which generates a different color light, while it has only basically two ionization states of much immediate interest to us radio folk.

GETTING CLOSER TO OHM

If you're like me, you had Ohm's Law drilled into your cranium and know that any electrical circuit needs to be, well, a circuit. You also know that you can't just have a battery out in space with nothing but a positive (or negative) terminal. So the concept of a totally isolated plasma, or a free-ranging electron cloud (the counterpart to a plasma), probably seems pretty alien. No need to hyperventilate; we haven't violated Ohm's Law, or any other electrical principle. But we have to address something that we usually ignore: the difference between local phenomena and *macro* phenomena. Let's dig into this a bit.

Most electrons come from somewhere, of course, and that somewhere is generally atoms. Most methods we have at our disposal for getting free electrons (or ions) require that we rip electrons from existing matter. It's irrelevant whether this matter is in the form of chemicals in a battery, or the copper wire in a generator. In our own little worlds, there are probably about as many electrons as there are protons, which is a good thing, or we'd have some scary electrical charges to deal with.

Any *practical* electron generator we can think of operates by separating two different types of charges. This means that any method you use to accumulate electrons on Item A results in fewer electrons on Item B, leaving Item B positively charged. Cats happen to be great static electric generators, if not especially practical ones. If you rubbed your furry friend, Van de Graaff the Cat, in the right way, you would achieve a highly charged cat. In general, rubbing Van de Graaff removes electrons from him, leaving you with a positively charged cat. This will leave the cat-rubbing object (your hand) negatively charged. This cat-rubbing process hasn't noticeably upset the equilibrium of the universe, though it probably upsets the equilibrium of Van de Graaff, perhaps in several ways. Now, if you let Van de Graaff the Cat outside, far beyond the "range" of your negatively charged cat-rubbing hand, he would still be positively charged.

The ohmically obsessed ham will immediately ask "positively charged compared to *what*?" The plasma physicist will say it's a pointless question. Charge is absolute. Van de Graaff the Cat has an absolute charge whether he remains in your living room or outside in the back yard. Though Van de Graaff the Cat's charge, compared to *your* charge, is irrelevant, his charge relative to the rest of the universe is *not* irrelevant. As he roams the back yard, he will pick up more dust than he normally would have if he had been tossed outside in an uncharged state. He will continue to accumulate dust until the charge of the dust particles (presumably negative) equal his initial charged state. He is now a somewhat dusty but discharged kitty.

INSIDE-OUT VACUUM TUBES

For many hams — at least, for most of us *older* hams — the most familiar form of plasma chamber is the vacuum tube (**Figure 12.2**). The primary difference between a vacuum tube and a vacuum chamber is a matter of *priorities*. Vacuum tubes work at very high vacuums compared to plasma chambers, because you generally don't want ions getting in the way of the electrons and clogging up the works. However, the concept of "free-ranging" or "one-sided" charges is exemplified in the vacuum tube. A vacuum tube is known as a *thermionic* device because it uses high temperature to ionize a hot filament. You could think of a hot filament as being a plasma in solid form. (This is really stretching the definition a bit, because a plasma is considered the *fourth state of matter*, as we learned earlier). Nevertheless, in a vacuum tube, electrons are liberated from the immediate influence of the filament (cathode) to become an autonomous, negatively charged electron cloud.

Although there are free-ranging charges in a vacuum tube, the

Figure 12.2 — A thermionic trio poses for the camera in all its glory. The venerable vacuum tube offers plenty of insight into the science of modern plasma physics. [Thomas Griffith, WL7HP, Silver Impressions, photo]

cathode exerts *some* influence on them. And actually, in exploring the vacuum tube, one can glean a real insight into why the ionosphere works so well. Let's give dogs equal time for our next example.

One image that comes to my mind when I think of a plasma or an electron cloud is that of a rather matronly woman with a yappy toy poodle on a long leash. The woman is free to travel at will, but at a somewhat less-than-lively pace, while the poodle can move around very quickly, but can't travel very far from its mistress. Now, imagine millions upon millions of these matrons, each with an associated poodle on the end of a leash. The matrons represent the ions and the poodles the free electrons. The electrons are free *up to a point*. If it weren't for the restraining influence of the matrons, the poodles would scatter to the far reaches of the solar system. However, because the electrons are *almost* free, they can accelerate easily in accordance with any electric field applied to them. This is a pretty fair picture of the ionosphere.

With a little modification of the above, we can also describe the vacuum tube. Imagine all the matrons tangled up and caught on a fire hydrant. Now our ion/matrons are more of a solid substance, like a tube's filament.

FLIES IN THE SOUP

One of the most curious (and sometimes frustrating) phenomena one encounters in a static plasma is a thing called *Debye shielding,* the distance over which the effect of a single charge's electric field extends in a plasma or electrolyte. It's a bit similar to skin effect in solid conductors in that there is a limit to how far an electric field can penetrate a plasma.

Let's imagine a plasma soup — that is, a spherical cloud of gas consisting of an equal number of free electrons and ions (it's never a pure electron or pure plasma cloud). If we now insert an ion into the middle of the soup, two things happen: the nearby electrons are attracted to the ion and the nearby ions are repelled. We now have a spherical region — at some fixed distance from our intruding ion — of higher ion density. Inside that sphere is a region of higher electron density. Already, we have created an ionospheric "layer" or "shell" of sorts (that is, a greater degree of separation between the ions and the electrons); however, *both* "shells" are much more conductive than the original soup.

Looking at just the ion shell for a moment, we see that we have built a sort of Faraday cage, a sort of barrier, around our original intruding ion. This Debye shielding prevents electric fields from penetrating it, but since fields are constituents of waves, this effect applies to incoming waves as well. There are only two things a wave can do when it encounters a wall: it can pass through it, or it can be reflected from it. Conservation of energy doesn't allow a whole lot of other choices. Fortunately for us radio amateurs, reflection is just what we want from the ionosphere.

But, of course, nothing in life is ever as straightforward as advertised, and plasma physics is no exception. Up to this point, we haven't talked much about absorption, which is a major factor in all radio-related phenomena, except in free space, where few of us radio amateurs reside or operate. Transmission and reflection of waves, or billiard balls as we saw earlier, actually *is* pretty straightforward — it's the fluffy felt lining on the pool table that puts a kink in the works. Calculation of radio absorption in a plasma is extremely difficult business, but one thing we can say is that absorption is always converted into heat. One of the cardinal rules of physics is that "If you want to know what is happening, follow the heat."

If you could actually measure the *heat* generated by a radio wave passing through the ionosphere, you could easily calculate the absorption losses. The problem is, where do you insert the thermometer in an ionosphere?

While we ponder that profound philosophical question, let's take a brief look at the actual *processes* of absorption. If you took high school physics, you might remember something about *elastic* and *inelastic collisions*. Basically, if you drop a marble on a concrete floor, it bounces nearly up to the same height from which it was dropped. This is a highly elastic collision. The proportion of initial height to which it bounces is called *the coefficient of restitution*, and for a glass marble on concrete (assuming the marble doesn't shatter), it's somewhere between 0.9 and 1.0. If you drop a blob of Play-Doh on the same floor, your result is quite different. Now the coefficient of restitution is about 0. Where has the energy gone? It's gone into heating up the Play-Doh.

In a way, the ionosphere is like a mixture of marbles and Play-Doh. An elastic collision between a radio wave and the ionosphere is a simple process: The radio wave accelerates an electron, the electron, reaching the end of its leash (remember our matrons?) then accelerates back toward its "owner," which in turn, creates a radio wave. Remember that electromagnetic waves are created by moving charges, and charges are moved by electromagnetic waves.

However, there are a few things involved in the process that are not perfectly elastic. First of all, the ions (matrons in our poodle example) are *not* infinitely massive, contrary to our initial impression. In the process of the electrons being accelerated, the ions to which they're associated are accelerated also. It's a small effect, but a *cumulative* one. The movement of the ions removes energy from the accelerated electrons. Where does this energy go? Some of it goes directly into thermal energy, as ions "rub" against each other. It's a crude analogy to friction, but a useful one. As with any other type of friction, this is converted into infrared electromagnetic energy. More interestingly, however, some of this energy goes into ion acoustic waves. These are primarily *compression* mode waves, much like sound waves, that carry energy away from the ions in all directions. Although the total energy diverted into ion acoustic waves is a rather small percentage of the total, it yields some of the more interesting results, as we will note shortly. Stay tuned.

WHAT'S THE ATTRACTION

Much Amateur Radio literature on the ionosphere, in my opinion, neglects magnetic effects in the ionospheric plasma and can grossly oversimplify the situation. Here, then, is a general description of how magnetic fields affect plasmas, especially in a more controlled environment, such as a plasma chamber. We can then extrapolate this behavior to the natural ionosphere, with a few caveats.

Let's return to our rudimentary plasma chamber for a moment. If we were to increase the voltage or decrease the spacing between the electrodes (or both), we would begin to see a significant current flow from the negative electrode to the positive electrode. The two plasmas would merge into one, but that new plasma would be in a rather agitated state. We would have rapidly accelerating electrons moving one direction, and ions moving the other. This situation makes for a very pretty, dancing purple arc, but it is a very difficult situation to analyze.

In the natural atmosphere, this situation would be called a lightning storm. Well, *almost.* Actually, lightning has a few complications added, because air at normal atmospheric pressure is extremely difficult to ionize: it takes about 20,000 V per centimeter to do the trick. It's actually a very good thing that air is such a lousy conductor — having highly conductive air would not be too conducive to human life, especially during a lightning storm. Because "normal" air is such a poor conductor, lightning occurs in a stepped process, where relatively small chunks of air are ionized in sequence, finally creating a sufficiently conductive plasma trail for the main lightning bolt to "fire." High-speed photography of lightning shows this process very clearly and it is indeed fascinating stuff to see.

What about pressure? Well, the lower the pressure is, the easier it is to ionize gases, up to a point. If there's no gas in your chamber at all, there's obviously nothing to ionize. If your desktop plasma chamber has a pretty good vacuum pump, you might find that your original semi-low-voltage plasma extinguishes after a while. This might suggest that there's a certain *degree* of vacuum that is ideal for getting a plasma going, as it were, and indeed that's correct. In fact, this happy medium between ease of ionization and "availability of stuff to ionize" is precisely why we have an electron density profile, as we showed in our section on ionograms. Plasma chambers also have an electron density profile, except in this case the profile is related to pump pressure, rather than *altitude*. It probably is fairly obvious that it's difficult to build a plasma chamber big enough to have an altitude pressure gradient. This is one of the primary limitations of plasma chambers, as opposed to the real ionosphere.

On the other hand, it's very easy to create a magnetic field with a *magnetic gradient* in a plasma chamber, and this is where things get really interesting. But before we can do this, we need to "de-agitate" our plasma.

ON Q

The term Q has great significance to most electronics and radio folk, or at least it should have. We use Q to designate the *quality factor* of an electrical component or circuit. However, in the plasma lab, Q is used to designate something quite different. It is a measure of the *quiescence* (lack of fluctuations) of a plasma. There are no actual units of Q in the plasma lab, but it's a very useful qualitative yardstick. In a plasma chamber, we want a plasma that's more like a warm glow than a lightning bolt, one that's uniform and isotropic; that is, it behaves the same in any direction — at least until we apply a magnetic field. To create a quiescent plasma, we use a curious device called, oddly enough, a *Q-machine*, which is fundamentally a light bulb, or an *unode* (single electrode tube).

Thomas Edison actually created a Q-machine without knowing it. He got annoyed by all those ions being boiled off the filaments of his embryonic electric lamps and clouding up the glass bulbs, so he put another electrode in there as an attempt to correct the situation (unsuccessfully), but in the process discovered the Edison Effect (the motion of free electrons through a vacuum) and then promptly abandoned the project because he couldn't see any way of making money at it. He left that to John Ambrose Fleming, who turned the Edison Effect into a practical diode, or Fleming valve.

In the modern plasma lab, we don't get annoyed by ions, as Edison did, but rather take advantage of them. To be fair, a few refinements had to be applied to Edison's light bulb to make a practical Q-machine, but in essence it's not much more than that common household fixture. A modern Q-machine takes a chunk of cesium or potassium (or other alkali metal) and "boils" it by means of a hot filament. These metallic vapors are then directed to a hot plate, made of tungsten, whereupon a more rapid and intense ion emission occurs, but without the sort of acceleration one has with an arc. We now have a chamber quite full of leisurely meandering ions, much as one might have in the ionosphere. Once we have a quiescent plasma, we can direct the ions from our Q-machine into another less cluttered chamber by means of simple plumbing.

Now, if we were to immerse an antenna in our quiet plasma and transmit an HF radio signal from it, there wouldn't be much difference from the same antenna in free space. (We'd obviously have to use a lot of inductive loading of the antenna to fit it into a practical plasma chamber.) If we were to put another receiving antenna in the plasma chamber, we would measure a slightly slower propagation velocity between the antennas than we would have in free space. This is because the dielectric constant of the plasma is greater than a vacuum. However, the behavior of the plasma would be absolutely uniform and isotropic — that is, the same in any direction. There would be no dispersion, refraction, or reflection from the plasma.

On the other hand, if we were to place one of the antennas outside of the plasma, things would change — dramatically. A signal transmitted from outside in a normal direction, that is, perpendicular to the walls of the chamber, would be partially reflected, and partially transmitted through the plasma. If the transmitted signal is not normal to the chamber wall, it will be bent, in much the same manner as the image of a straw in a glass of water appears bent below the surface of the water.

This process is fairly non-dispersive; in other words, the amount of bending is relatively independent of wavelength. We have essentially just two refractive indices: that of air outside the chamber and that of the plasma inside the chamber. The refraction takes place quite suddenly at the demarcation between the two media.

Here we have a profound difference between the plasma chamber and the ionosphere. In the plasma chamber, the quiescent plasma has the same refractive index throughout. In the ionosphere, the plasma has a refractive index that continually changes with respect to altitude. This is the primary reason for the ionosphere being "frequency dependent."

By the way, the reason light is dispersed in a glass prism is because optical glass itself is frequency dispersive. This is actually a somewhat peculiar property as far as most solid materials go. We only think this is normal because glass is so common. Likewise, water itself is a really bizarre substance, but we consider it normal because there's so much of it.

Regardless of the process on the atomic level, the behavior of the ionosphere is very similar to familiar optics principles of reflection, refraction, dispersion, and so on. For the most part, it's simply a matter of scale that sets the ionosphere apart from lenses and prisms and such.

OUT OF UNIFORM

But if we apply a magnetic field to our otherwise uniform, quiescent plasma, things change dramatically — and get a lot more interesting. First of all, we no longer have a uniform plasma: we now have a plasma with a *density gradient* that is proportional to the magnetic field. In other words, we now have a highly *dispersive* media, and that dispersion can change radically in a very small space. Second, the dispersion or density gradient is dependent on whether you follow along *parallel* to the magnetic field lines or *perpendicular* to them. You now have actually *two* refractive indices, one being greater than the quiescent plasma and one being lower, depending on which direction the signal is traveling. This property is known as *birefringence*, and it is primarily responsible for the *non-reciprocal* behavior of ionospheric radio propagation, contrary to the common wisdom that the ionosphere is reciprocal Not only is the velocity of propagation different, depending on which direction the wave is passing through the plasma, but the amount and *direction* of dispersion is dependent on that also. To add insult to injury, in the actual ionosphere, not only do you have the density gradient due to the altitude changes, but you also have all these magnetic properties thrown in for good measure. Is it any wonder that predicting propagation is a bit tricky?

PUMPED UP

Now that we've described a fairly *normal* plasma, let's talk about a few oddities one may encounter at some time in one's ham radio career.

If you've operated on "low bands" for any significant time at all, you have probably encountered signals that seem to be far stronger than they should be. This is a pleasant, but baffling, surprise when it happens. It's most likely to occur on 80 meters, or at least it's more frequently noticed on that band. It usually occurs late at night over the "dark path" when the ionosphere has retreated to a high altitude and the absorption is very low. Chordal hop propagation is well known to exist; this is when a signal can bounce several times off the ionosphere without any intervening ground reflections. Naturally this type of propagation path would be much less lossy than a normal ground-sky-ground-sky-ground-sky path. However, even when chordal hop is accounted for, signals can be *much* stronger than can even be accounted for by free space loss! And this isn't just someone's imagination; actual field strength measurements have been done that confirm less than free space losses of such signals. How can this be?

The answer to this is most likely parametric amplification. We are careful to say most likely because the jury is still out on this. However, when such a phenomenon can be created consistently in a plasma chamber, it's a fair bet that something similar is happening "in the wild." Most microwave aficionados have encountered a parametric amplifier, or paramp, at some time. This unusual device doesn't use variable resistances to amplify, as a transistor might, but rather uses a *variable reactance* to achieve amplification. Typically, a *varactor diode* is used at the core of a parametric amplifier, which is pumped by an intermediate frequency. The amplification process is lossless, compared to a lossy transistor, but please don't get the impression that we're getting something for nothing. Although the paramp can be difficult to describe electrically, a very common mechanical analogy gives us a pretty clear picture.

If you were a fairly coordinated child (or at least knew one), you probably figured out how to get a swing going without someone giving you a push. You probably did this by standing on the seat, grabbing the chains and writhing in a very scientific fashion. This process seems to violate one of Newton's laws or something, since there is nothing for you to push against. And yet we all know it works. Actually, this does require a minuscule amount of friction of the chain against the hinges — a perfectly frictionless system would need a minuscule offset nudge to get started, regardless of the child's skill. However, once it is going, you can add energy very quickly by contorting at the right time in the right direction. This is parametric amplification.

But we have a guiding principle in everything we do in physics: NFL, for no free lunch. We want to be extremely careful to emphasize that there is nothing even remotely resembling perpetual motion with regard to parametric amplification, either in the plasma chamber or in the natural ionosphere. In a parametric amplifier, you never get more power out than you put in. However, the source of the input power may not be as obvious as it might be in a conventional amplifier.

In a normal RF amplifier, for instance, the input power all comes through the dc plate or collector supply. In a microwave paramp, the primary driving power is a "pump" oscillator at a *different frequency*. This is the key — and the mystery — of parametric amplifiers. Just as the young you probably didn't know exactly *how* you got the swing going, it can be a bit tricky to explain how a signal at frequency B can amplify a signal at frequency A. To put it in very simple terms, the slope of the pump frequency signal only has to coincide with the slope of the amplified frequency signal periodically, and as long as these occasional "coincidences" are in phase, energy from the pump can be transferred to the desired frequency.

In the plasma chamber, we can use ion acoustic waves to amplify radio frequency waves. The ion acoustic waves have no direct correlation to the radio waves, but if we allow the peaks of the ion acoustic waves to coincide with those if the RF wave often enough, we can obtain amplification.

The only missing ingredient is the energy to create the ion acoustic waves in the first place. This can actually come from several sources, none of them (necessarily) even remotely related to the incoming RF frequency. This can come from plasma *or cyclotron resonances,* the natural rotation of ions around a magnetic field line. (This is the primary means used to create parametric amplification in a plasma chamber). It can come from a secondary RF pump frequency, as we saw with the Luxembourg Effect.

It's an intriguing process, because not only will the Luxembourg Effect give you the difference frequency between two input frequencies, but if you re-mix the difference frequency into one of the original signals, you get the original frequency back too. This is one of the more likely processes of "in the wild" parametric amplification. In fact, the Earth's ionosphere has all the ingredients necessary for parametric amplification: a magnetic field, a supply of ions, plenty of available "pump" energy, both electromagnetic and ion acoustic. The only question is, how often do all the ingredients come together at the same time to be useful? Is parametric amplification in the wild a long shot? We don't know. But it's certainly not out of the realm of possibility. No more so than our mysterious LDEs.

A MATTER OF SCALE

One of the nice things about working with antennas is that they scale perfectly. One can model (or build) an antenna for one frequency, and then change all the dimensions proportionally and it will work precisely the same at some other frequency. Electromagnetism is extremely linear in this regard — at least to the best of our current knowledge. Unfortunately, plasmas do not scale so nicely. We have certain fundamental values like ion and electron cyclotron frequencies and such which can be scaled only if we change the size or mass of ions or electrons, and we haven't been able to do that. Such parameters set absolute limits on ionospheric parameters as varied as critical frequency or MUF to D-layer absorption and other related phenomena. Unlike electricity, plasmas are compressible, and this in itself adds orders of magnitude to the complexity of the matter.

Does this make ionospheric study a hopelessly complicated cause? Not at all; it just adds more to the intrigue, and suggests countless unexplored nooks and crannies to investigate, most of which are accessible to anyone with a radio and an antenna. And a functional set of curiosity glands.

13 Smith Charts, Scattering Parameters, and Sundry Science Tools

Few things elicit more fear and loathing in the heart of the ham than the Smith Chart. But we will show that the Smith Chart, in concert with some other related tools, is actually a major improvement over the way we radio folk used to perform difficult transmission line calculations. Moreover, studying transmission line theory gives us a tremendous insight into radio propagation itself. It also gives us a satisfying and elegant way to look at more conventional circuitry. It is well worth your time to become comfortable with the Smith Chart.

COMPONENTS AND LUMPED CONSTANTS

When dealing with conventional low-frequency electronics, we can generally consider any component a *lumped constant.* This means that all the properties of a component we're interested in exist at a specific point in space. A resistor, for instance, has all its resistance at one location; a capacitor has all its capacitance in one place; an inductor has all its inductance in one place, etc. When a component is sufficiently small in relation to the wavelength of the signal it is subjected to, we can consider the component a lumped constant. There is no strict line of demarcation between a component being a lumped constant and being something else, but it's safe to say that if a component is on the order of 0.001 or less of the wavelength, it is essentially a lumped constant.

A circuit with lumped constants is relatively simple to analyze. For one thing, we don't have to consider the propagation time of a signal or wave passing through the component. If we apply 1 V dc to one end of a wire, we can assume that voltage appears at the far end of the wire instantaneously. Even if a lumped constant has a great deal of reactance, we can still assume that the relative voltage and current phasing is the same at one "end" of the component as it is at the other. Such circuits can be described entirely by their *time domain* response; that is, a simple oscilloscope can tell you all you need to know. Even if we should need to know the *frequency response* of the component, we can get that information entirely from the time domain response with the use of appropriate mathematical transformations.

Most circuit analysis methods assume the existence of lumped constants; in fact, the very definition of circuit is based on some wiring of lumped constants. Ohm's Law, Kirchhoff's Current Law, and Kirchhoff's Voltage law themselves are based on lumped constants. But such analysis becomes inadequate when we have *distributed systems*, the primary examples of these being antennas and transmission lines.

As fast as the speed of light (or any other electromagnetic radiation) may seem, at *only* around 300,000,000 meters per second, it is far from being *infinitely* fast. The limitations of the speed of light become very apparent once we begin to approach frequencies even remotely resembling normal radio frequencies.

When the physical size of electronic components becomes significant, we can no longer assume that an action at Point A causes an immediate result at Point B. We now have to consider the propagation time between Point A and Point B, as well as a whole new category of interesting effects. Our circuit elements become distributed elements, in which the properties of resistance, inductance and capacitance are stretched over a significant distance. In such instances, it's far more useful to study the *wave interaction* in such components than the simple electrical responses. Phenomena such as reflection, refraction, absorption, and interference become all important, just as they are in radio waves in space.

Fortunately, and perhaps surprisingly, the study of wave behavior in circuits in no way violates conventional electrical circuit principles; it actually reinforces these principles, but from a different point of view.

LADDER LINE TO ETERNITY

We mentioned earlier that there's no strict line of demarcation between lumped constants and distributed components, and that as long as a component is very small compared to a wavelength, it can be considered a true lumped constant. However, as we now know, it's best to study the ideal first, so let's start with an infinitely long transmission line, in which there is absolutely no lumped constant behavior to confuse the issue.

The simplest transmission line is constructed of two parallel wires. In the early days of radio, such conductors were held apart by uniformly spaced insulators, and became known as ladder line. True ladder line is hard to obtain these days, and most hams use a somewhat inferior substitute, such as twin lead or window line. Some hams still build their own ladder line because of its superior performance and flexibility.

An ideal transmission line consisting of two parallel conductors of infinite length, our "Ladder Line to Eternity," has some unique and important properties. If we were to connect a battery to one end of the transmission line, a surge of dc current would travel toward the far end at near the speed of light. Now, the astute observer might ask, "How can a surge of *current* even exist? It's an open circuit, and you can't have any current in an open circuit, no matter how much voltage you apply." Well, this is where we have one of the profound differences between a lumped constant and a distributed circuit. We have, throughout the length of the line, a distributed capacitance — the capacitance between the two conductors. As we know from circuit theory, applying a dc voltage to a capacitor will indeed cause current to flow — at least instantaneously. If there's some resistance involved, the current will follow the RC time constant pattern: it will drop off exponentially with time as the capacitor charges toward the applied voltage.

However, it's not *quite* so simple. Since the "capacitors" in this case are not plates, but rather long wires, the charge will not occur simultaneously along its length. In fact, the charge can only "spread" along the length of the capacitor at the speed of light. In effect, we have an infinite number of infinitesimally small capacitors, each charging sequentially along the length of the transmission line.

But that's not all. Each incremental length of transmission line also has inductance. In fact, we have the equivalent of an infinite number of infinitesimally small inductors in series along the line, interspersed with our infinite number of parallel capacitors. Just as in the case of normal inductors, the incremental inductance tends to limit changes in current flow, in opposition to the applied voltage. The composite results of all these incremental capacitors and inductors gives us the all-important *characteristic impedance* of a transmission line, which is designated Z_0.

The formula for the characteristic impedance of a uniform, parallel transmission line is:

$$Z_0 = 276 \log_{10} 2S/d \hspace{4cm} \text{(Eq 13.1)}$$

where
   S is the center-to-center spacing between the conductors
   d is the diameter of the conductors.

As with "normal" circuits, Z is expressed in ohms. In reality, Z is a complex number even for an ideal transmission line, but for the most part we can ignore the imaginary part of the transmission line itself, as we usually do when using the Smith Chart, as demonstrated later.

So, now that we have an actual characteristic impedance on our transmission line, measured in real ohms, the idea of current flowing in an open circuit begins to make some sense, at least mathematically. Now here's the amazing part, though admittedly, a bit difficult to prove. A battery connected to an infinite length, open-ended transmission line will produce a continuous dc current. The current will be precisely the battery voltage divided by the characteristic impedance of the transmission line: I = E/Z, our standard Ohm's Law for ac formula. Again, this would be rather difficult to demonstrate, since your friendly local electronics shop probably doesn't sell ladder line in infinite length spools.

On the other hand, it's quite easy to demonstrate transmission line principles on much shorter than infinite transmission lines, if one uses short *bursts* of energy. Let's take a piece of transmission line 500 feet long or so. Now, for strictly logistical reasons, we will use *coaxial* transmission lines for this sort of demonstration. Coaxial transmission lines are not fundamentally different from balanced or parallel line transmission, except that the characteristic impedance tends to be much lower. One nice thing about coaxial transmission lines is that you can roll up great lengths of the stuff without shorting it out, which would, obviously, be a bit of a trick with uninsulated ladder line.

Because a coaxial transmission line (at least the flexible kind you can roll up) has an insulator between the conductors that is something other than air, the velocity of propagation is somewhat less than the speed of light. Practical transmission lines can have a velocity factor as low as 66%, which makes this sort of demonstration even easier. A general rule of thumb is that light travels about 1 foot/ns. In flexible coax, radio waves travel about 8 inches/ns (0.66 times 12 inches), so it takes about 750 ns for a radio wave to travel the length of our 500 foot roll of coax.

Let's fire up the trusty old dual trace oscilloscope. We'll connect one end of the coax cable to Channel 1 of the oscilloscope through a T connector and the other end of the cable to Channel 2 with another T connector. We'll put a 50 Ω termination resistor on the second port of this T connector, or if you have a fancy oscilloscope with internal terminations, you can use that. Now, obtain a pulse generator that can produce 10 ns pulses with a repetition rate of 100 kHz. Actually, you can build one of these with about three dollars' worth of parts, using a 555 timer chip and a handful of other parts. Now we connect the pulse generator to the open

T port on Channel 1 of the oscilloscope.

With the oscilloscope in the dual trace mode (preferably "chop" mode), set the trigger to Channel 1. The pulse displayed on Channel 2 should be delayed by about 750 ns. The pulses on Channel 1 and Channel 2 should be very nearly equal in height.

Now, if you replace the termination resistor with a dead short, some interesting things will happen. The most obvious will be that the Channel 2 signal disappears. But, more importantly, you will see *inverted* pulses on Channel 1, interspersed with the outgoing pulses. These inverted pulses are *total reflections* from the short circuit at the end of the transmission line.

If you remove the short circuit, leaving the transmission line unterminated, you will also see some interesting things. First of all, the pulse at Channel 2 should be twice its original height. Second, the reflected pulse displayed on Channel 1 will be non-inverted. What we have demonstrated is *time domain reflectometry* (TDR), which is a commonly used tool for finding faults in long transmission lines. This basic method works on anything from power transmission lines to fiber optics. In practical applications, one doesn't have access to the far end of the cable, so it's difficult to plug that into an oscilloscope. However, any reflection will show up on Channel 1.

We can learn a lot about transmission lines this way. First of all, we know that a properly terminated transmission line has no reflections. If the termination resistance is lower impedance than the Z_0 of the transmission line, the reflection is inverted, as in the case of a short circuit. If the termination is higher impedance than Z_0, the reflection is non-inverted. And we can determine the distance to the termination (or any other fault) by measuring the time, and multiplying it by two, to account for the round trip. This principle is very much like closed circuit radar.

Using short pulses in this manner, it's simple to isolate the forward going pulse from the reflected pulse. However, most of what we do in Amateur Radio, as well as a good deal of radio science, consists of continuous streams of radio frequency energy — at least continuous in terms of normal transmission line lengths. We can no longer separate the forward from the reflected pulses. Which brings us to the concept of wave interference, which leads us to the Smith Chart, which we know you are all champing at the bit to explore in all its glory.

BASICS OF SMITH CHARTS

Most texts on this topic present a treatise on wave and reflection mechanics first, and then show how to use the Smith Chart to do the calculations. I find it much more useful to present the Smith Chart right off the bat, and then use it to help explain the mechanics. I am confident you will appreciate this approach.

Figure 13.1 shows a basic Smith Chart (we'll be adding more detailed features to this template as we progress). This chart was created by *Origin 8.6*, but with a lot of the gory details left out, so it meets my fat purple crayon standard of simple. It has very coarse resolution as you can see, but it's extremely useful for getting accustomed to the Smith Chart in a painless manner.

Since this template was created with *Origin*, it's only appropriate that we start at the origin, which is the center of the circle. Unlike most graphs with which you're probably familiar, the origin of this graph isn't 0,0 but rather 1,0. What does this mean? Well, within the mathematics world, we sometimes refer to 1 as the *identity* function. It is used as a comparison for any number of subsequent operations. In the Smith Chart, the 1 is the identity for the *characteristic impedance*, whatever that may happen to be. Typically, this is 50 Ω, which is what the majority of coaxial transmission lines are in radio frequency work. However, this chart can be used for transmission lines of any characteristic impedance. You can also obtain *normalized* Smith Charts for specific characteristic impedance lines, which require somewhat fewer steps to use, but it's best to learn how to *normalize* impedances yourself, using a universal Smith Chart, such as this.

The reason this works is that all Smith Chart calculations are based on the *ratio* of characteristic impedance to some other value or values; that is the ratio of the characteristic impedance to the *localized* impedance at some point on the transmission line. Let's look at the horizontal "crossbeam" to start with. Here we have a range of resistance values ranging from 0 on the left to infinity (∞) on the right. Any value on this horizontal bar will be purely resistive. Notice that the values are not linear from the left extreme to the right, but are rather geometrically proportioned.

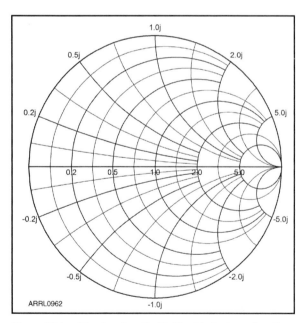

Figure 13.1 — The daunting Smith Chart is far less daunting than the mathematics behind it. In fact, the Smith Chart is a surprisingly simple and intuitive tool for the radio scientist and radio practitioner.

If we normalize this chart to 50 Ω, that means the center is 50 Ω. To find the normalized value of any point labeled on this horizontal, we simply multiply that value by 50 as well, so our reference values are: 0, 10, 25, 50, 100, 250, and ∞.

Notice that each of these values lies on a *circle*, and each circle is tangent at infinity. These circles represent the *resistive component* of the impedance that lies on that circle; however it is only where the circle intersects the horizontal that the impedance is purely resistive.

Next, we have the *inductive reactance* arcs, which lie above the horizontal bar, but are asymptotic to it. These all terminate on the outside "infinity circle" at $+j$ increments, again geometrically incremented. These are also normalized values. If the chart is normalized to 50 Ω, we need to normalize the $+j$ values to 50, as well, so we have, moving clockwise: $j0$, $+j10$, $+j25$, $+j50$, $+j100$, $+j250$, and $+j\infty$. Any value lying on one of these arcs will have the reactance value indicated on the outer infinity circle.

Mr. J., the Smooth Operator

The term "j operator" can be a bit confusing to the newcomer, but a firm grasp of the concept will make the Smith Chart a snap to understand. The important thing to know is the *physical significance* of the j operator: it always represents *reactance*. The lack of j can mean only one of two things: either the circuit is *purely resistive*, or the circuit is *resonant*.

One of the slick things about the Smith Chart is that you can immediately tell if a particular value of impedance is purely resistive or resonant. The point will likely be somewhere on the horizontal bar if it is a pure resistance. Anything on the horizontal bar is going to be $j0$ Ω (this is true for any normalized impedance, by the way).

One of the amazing (or disturbing) properties of a transmission line is that it can convert a purely resistive impedance to one containing reactance. In fact, this is the case whenever there is a mismatch between the transmission line and the load, even if the load is perfectly resistive; the SWR circles we'll discuss show this. If fact, for any mismatched load, there are only two points where the input impedance is purely resistive: where the circle crosses the $j0$ bar.

One other important feature of the j-operator is that you can always cancel out any value of j by an equal and opposite j. For example, if you have $+j25$ Ω (inductive reactance), you can always add $-j25$ Ω to that (capacitive reactance) and bring the impedance down to the "crossbeam."

We should probably mention here that there is nothing particularly sacred about resonance in general, other than that resonant circuits are easier to calculate. In many cases, eliminating j, either mathematically or physically, can make a rather unwieldy RF problem a simple Ohm's Law problem. Again, the Smith Chart comes to the rescue by presenting, in a simple graphic way, things that can be calculated only by extremely complex hyperbolic trigonometry.

Finally, *below* the crossbeam, we have *capacitive reactance* values. These are patterned exactly as the inductive values, but with a –*j* operator (see "Mr. J., the Smooth Operator" sidebar).

GOING TO EXTREMES WITH SMITH CHARTS

Sometimes, when demonstrating a physical phenomenon, it's best to show the extreme cases and work back toward "normalcy" from there. The Smith Chart is a good place for this practice, for reasons you'll see shortly.

In learning to use the Smith Chart, it's always best to start with a known value of impedance at the end of a transmission line and work back toward the generator. This is by no means the only way to work a Smith Chart, but it does help emphasize some principles that will serve us well from here on out. Plotting what we did in our TDR oscilloscope demonstration, we get **Figure 13.2**. Notice the small square block on the left end of the horizontal bar. This is 0+*j*0 Ω, a dead short.

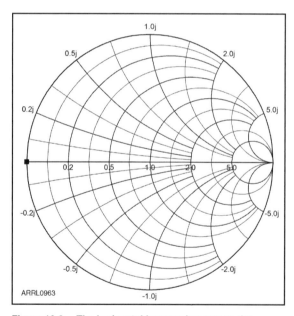

Figure 13.2 —The horizontal bar running across the middle of the Smith Chart represents the resistive component of an impedance. The intersection of the circles and the "crossbeam" represent resistance values, in this case, 0.2, 0.5, 1.0, 2.0, 5.0, and ∞, moving from left to right. These are normalized values, meaning that each one needs to be multiplied by the characteristic impedance of the transmission line in question (most commonly 50 Ω).

Now, on a prefabricated Smith Chart, you will have notation around the outer ring. Going clockwise around the perimeter, you will have "Wavelengths toward Load," and going counterclockwise "Wavelengths toward Generator." There might be more than one scale here, but generally it will be in degrees. Notice there are only 180° around the entire perimeter. This is because the impedance of a transmission line repeats itself every ½ λ, or every 180° electrical. This alone gives one great insight into transmission lines. The input impedance of a transmission line that's an exact multiple of a ½ λ will be the same as the impedance at the load end. Conversely, a transmission line that's a ¼ λ long *inverts* the impedance. On the Smith Chart, a ¼ λ transmission line's input impedance is *diametrically opposed* to the load impedance.

In the extreme case of a dead short, we'll see an open circuit at the input end. If the load is open, we'll see what

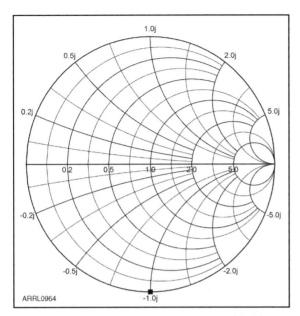

Figure 13.3 — This plot shows an impedance of 0–j50 Ω. The resistance is 0 because it lies on the zero resistance circle, the outer circumference and the –j1 arc, which intersects that circle (the reactance values are also normalized to 50 Ω). Notice that the outer circle also intersects the "crossbeam" at infinite resistance.

appears to be a dead short at the input end of the transmission line. Keep in mind that you will not read a dead short with an ohmmeter; this appears only at the frequency in question. It must be measured with a bridge of some sort, or if you've got access, a network analyzer (we'll talk about network analyzers shortly).

Here are a few more things to note about this particular point (the square block). If we were to scoot this point around the perimeter of the chart, the impedance would always be purely reactive. Going counterclockwise, it would be capacitive, and clockwise, it would be inductive. If we follow the chart counterclockwise ⅛ λ, we'll find that we are at $-j1$ Ω (**Figure 13.3**). But keep in mind we have to *normalize* this impedance first, which means we multiply both complex components by the characteristic impedance of the cable, in this case 50 Ω. So our new point is actually $-j50$ Ω. This tells us that a transmission line with a dead short on the far end will look like a pure capacitor if it's between 0 and ¼ λ. You can actually make fine capacitors out of sections of transmission lines. With a shorted transmission line, there will only be two places where the impedance has any *real* value: 0 Ω or infinity Ω. (It might be a bit of a philosophical stretch to call infinite ohms *real*, but dimensionally it works fine.)

If we lengthen our transmission line ¼ wave from our last point, we will find ourselves at top dead center of the chart. This will be $+j50$ Ω, again remembering to normalize (**Figure 13.4**). This will be a pure inductance.

Now, if you're an active radio amateur, you probably won't be transmitting into dead shorts (at least not intentionally). Most actual antennas have a combination of resistance and reactance. In the case of a perfectly matched antenna (or dummy load), the impedance of the transmission line will be 50 Ω no matter how long it is, which makes the Smith Chart entirely unnecessary. But let's look at a real antenna. If we're either rich or well-connected, we can use a *vector network analyzer* (VNA) to measure the impedance of an antenna. A VNA looks sort of like an oscilloscope, but it has a Smith Chart on the screen. Let's say we connect our VNA to a

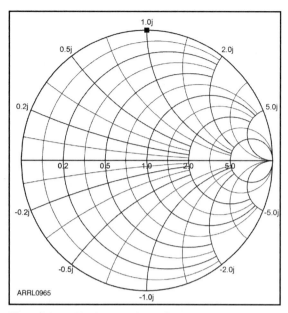

Figure 13.4 — Moving exactly ¼ λ from the previous graph (counterclockwise) we find our plot at +1J. This is a pure inductive reactance of 50 Ω, after normalization. Each "lap" around the Smith Chart represents ½ λ along a transmission line, counterclockwise moving away from the load, and clockwise moving toward the load.

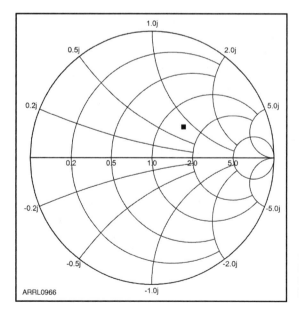

semi-sort-of-resonant 40 meter dipole. We come up with an impedance of 72 + j40 Ω. To plot this on our "denormalized" Smith Chart (since, alas, we do not include a VNA with this book), we need to divide both components by our characteristic impedance, in this case 50 Ω. So we have 72/50 = 1.44 Ω, our resistive component. Our reactance is 40/50 or 0.8 Ω inductive. Let's plot this denormalized point on the Smith Chart (**Figure 13.5**).

Well, now we see this point sort of hanging out in space, and we really don't have enough grids to do it justice, so we will have to embellish our "fat purple crayon" Smith Chart by adding a few more lines (**Figure 13.6**). Not to worry; everything will be exactly the same, just with a bit finer "grain." Don't be confused by the "dead spots" near the right end of the chart. These are left open because otherwise it gets really crowded out there. But take a look at our plotted point. You want to find the nearest line that crosses the horizontal crossbeam and the closest line that intersects the outer ring.

Let's do something especially tricky. Take a drawing compass and poke the point in the center of the chart, the origin, and place the pencil tip on our 1.44 +j0.8 point. Draw a complete circle around the origin, intersecting our impedance point (**Figure 13.7**). Well, what do we have here? This circle shows all the possible impedances that can exist on

Figure 13.5 — Here we have plotted an impedance that has both resistance and reactance (see text). We know it has reactance because it is not on the "crossbeam."

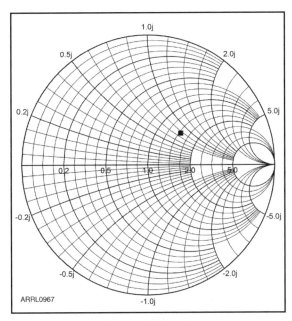

Figure 13.6 — This is the same plot as Figure 13.5, but with more grid lines to more accurately locate the plotted point.

a transmission line with the designated load impedance. Notice that while there is an infinite number of values it *can* have, there is also an infinite number of values it *cannot* have. We're more interested in the impedances that it *can* have.

This circle represents the *standing wave ratio* (SWR) on the transmission line; more specifically, the *radius* of the circle is the SWR. A transmission line that is perfectly matched (1:1 SWR) will appear as nothing but a dot at the origin. A transmission line with *infinite* SWR will appear as a circle around the outside perimeter. Most real loads are somewhere in between. Prefabricated Smith Charts will usually have a linear graph labeled "radially scaled parameters," which will allow you to read the SWR directly by projecting the radius of the circle to the linear scale.

Take another look at our SWR circle. We see that, obviously, the circle crosses the horizontal axis at just two points, one at about 24 Ω, and one at about 104 Ω. We can normalize this for you, as long as you recognize this is cheating (**Figure 13.8**). We've also de-cluttered it a bit so we can concentrate on the important parts again.

For some purely practical reasons, it's often easier to match a non-reactive (purely resistive) load to a transmitter,

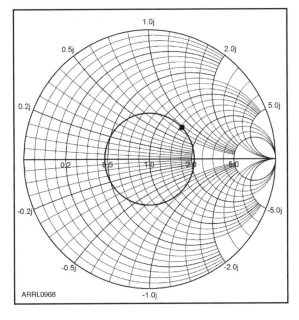

Figure 13.7 — If we draw a circle passing through our complex impedance point (see text), centered at the middle of the Smith Chart, we now have the locus of every possible impedance that can exist on our transmission line.

Smith Charts, Scattering Parameters, and Sundry Science Tools

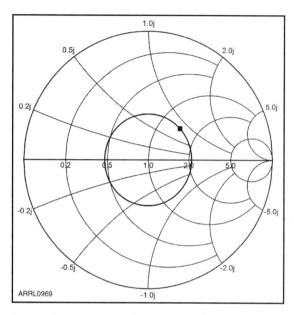

Figure 13.8 — Here we have removed the fine-grained grid marks for clarity, but the impedance and SWR are identical. Notice that the SWR circle crosses the "crossbeam" at two points: one at just under .5 Ω and one at just over 2 Ω (normalized to about 25 and 100 Ω, respectively).

even if the resistive part is a bit off. The neat thing here is that if we look at these two resistive points on the horizontal bar and we divide either one by the characteristic impedance, 24/50 or 104/50, and then take the reciprocal of any value less than 1, we will have our SWR. In this case, it's about 2.08:1. SWR will always be equal to or greater than 1:1.

How do we get from our original point to either 24 or 104 Ω? First draw a line segment from the origin, through your plotted point, intersecting the outer infinite SWR ring. Take a protractor and measure the angle between this new segment and the horizontal bar. The shortest path will be counterclockwise about 150°. But remember the Smith Chart repeats itself every 180° (the whole circumference of the chart is 180°, not 360°). Therefore, each degree you measure really equates to a half degree of feed line length, so you'll add about 75 electrical degrees (half of 150) of coax. This will give you an input impedance of 24 Ω, purely resistive. Almost any transmitter can handle that.

Both the *ARRL Handbook* and the *ARRL Antenna Book* have several more Smith Chart examples for a variety of situations. We recommend doing the exercises, as it will give you a real insight into wave mechanics and such.

S-PARAMETERS

Closely associated with the VNA are measurements called *scattering parameters*, or S-parameters. However, even without fancy equipment, you can use S-parameters to describe any number of complex circuits. Although S-parameters are used to describe basically distributed networks, such as transmission lines, they are perfectly suitable for lumped constants as well. In fact, the amazing symmetry between circuit theory and wave theory is so revealing that everyone involved in electronics should understand S-parameters. They're beautiful things.

S-parameters are defined in terms of a stimulus and a response to that stimulus; they are described by a port number for the stimulus and a port number for the response. However, by convention, the response is given first and the stimulus second. For folks familiar with C programming, this is actually quite intuitive (a C program gives you a "return" in response to an input, or the MAIN function).

There are two configurations hams will most often encounter when working with S-parameters: the 1-port and the 2-port. The 1-port is generally associated with an SWR bridge or an antenna analyzer, meaning that we're looking at something that's at the "end of the road," such as an antenna. An SWR meter stimulates the antenna and looks at its response (the impedance) at the same set of terminals. So SWR is defined in S-parameter notation as S_{11}. The first subscript defines the reflected power (response), and the second subscript defines the forward power (the stimulus). Obviously, since there is only one port (the load), you don't have a lot of choices in the subscripts.

Sometimes you will want to look at what's happening at both the input and the output of some device, such as a filter or an amplifier. In this case, you would want the 2-port model. Any 2-port device has four possible S-parameters. In this case, 1 is the input port and 2 is the output port. Here are the four S parameters: S_{11}, S_{12}, S_{21}, and S_{22}.

Using our response/stimulus definitions, we see the first parameter defines reflected power from Port 1 in response to forward power into Port 1; the second parameter is power out of Port 1 in response to power into Port 2; the third is power out of Port 2 in response to power into Port 1; and the last is reflected power out of Port 2 in response to power into Port 2.

Let's take the example of a low-pass filter. In general, we would be most interested in the S_{21} parameter; that is, what's coming out of the output of the filter when we put something into the input of the filter. If the filter is perfectly terminated, this parameter may be the only one we need. But what if the termination of the filter is not perfect? The mismatched termination itself will reflect energy back into Port 2, which may have an effect on Port 1. In this case, we might also want to see the S_{12} parameter. And, of course, if the input impedance of the filter is not what we expect, we might be interested in the S_{11} information, too. In fact, probably the only parameter we can ignore in a passive filter is the S_{22}.

S-parameters are particularly useful in describing amplifiers. Theoretically an amplifier should be unidirectional, or unilateral. All kinds of amplifier ills, including instability and feedback, can occur if an amplifier is not unilateral. S-parameters are a convenient way of describing and detecting non-unilateral behavior, and can help one remedy that situation (sometimes). Oscillators are amplifiers that are intentionally non-unilateral, and S-parameters can help you with those, too.

Unilateralization of components and circuits can also greatly improve noise performance, and S-parameters can quickly reveal some tough noise problems that just don't show up by more conventional notation or analysis.

If you should be privileged enough to get your hands on a good VNA, use caution. These are expensive and somewhat delicate devices.

We trust that this chapter on the Smith Chart has taken some of the "daunt" out of a sometimes daunting subject. As with any other unfamiliar subject, there's no substitute for just diving in and working out some problems. You'll be glad you did.

Put SPICE in Your Life for Circuit Simulation

One of the most powerful (and earliest) computer programs available for the radio amateur is *SPICE (Simulation Program with Integrated Circuit Emphasis),* an open-source program with a long and circuitous (pun intended) history. *SPICE* is now available in myriad versions, all with a common engine doing the actual number crunching. The name may give the impression that the program is good only for modeling complex integrated circuits, and while this was the original driving force behind its development, *SPICE* is also one of the best tools available for designing and analyzing conventional radio frequency circuits. Let's examine this granddaddy of electronic circuit modeling, first by more carefully defining what modeling actually is and a few things it isn't.

Most of our understanding of the universe is based on a model of some kind. Whether that model is a tiny plastic car or the atom, it represents reality, but it is not reality *per se*. It certainly leaves a lot to be desired. But although incomplete and not perfectly accurate, a model should be readily recognizable. Here's a case in point.

Our typical concept of the atom — to use a fruit model — consists of an orange (a nucleus) with a few grapes (electrons) circling around it a few grape-widths away. In reality, if we were to accurately proportion a hydrogen atom, if a proton were the size of an orange, the electron would be the head of a pin 3000 miles away.* This widely accepted model of an atom utterly fails to portray the vast amount of empty space there is inside an atom. However, for most of our understanding of electronics, the

---

*While modern understanding of the electron tells us it actually has no size whatsoever, it does have a finite "personal space" which allows us to arrive at this approximate proportion.

orange and grape model works quite well. On the other hand, observation of more complex behavior requires more accurate scaling of the atom.

One of the goals of science is to develop ever more accurate models of the universe, knowing full well there will never be a perfect model. Once we think we have a perfect model (such as a simple trinity of electrons, protons, and neutrons), along will come some wise guy who proposes things like quarks, gluons, neutrinos and positrons, upsetting our neat, blister-packaged model. The problem with these wise guys is that they're frequently right, so we have to adjust our models to agree with new discoveries — which brings us back to *SPICE* circuit modeling.

SPICE circuit modeling makes a few assumptions based on an incomplete but useful model of reality, primarily being that "pure" components such as perfect capacitors, inductors, and resistors exist.

Anybody who's wound a coil for a crystal radio knows that a perfect inductor is a fabrication of the mind. You can easily measure the resistance of a "perfect" crystal radio coil with an ohmmeter. Conversely, as you move into the VHF/UHF frequency region, you find that most real resistors have as much inductance in the leads as they have resistance (unless you use specially designed resistors for this purpose). Likewise, most real capacitors have lead inductance and equivalent series resistance that can put a kink in the works.

Now, depending on the sophistication of your particular version of *SPICE*, these "parasitic" values of components are compensated for, sometimes quite accurately. However, the basic *SPICE* engine underneath all the bells and whistles makes its calculations based on fundamentally perfect components.

SPICE NETLIST

The common entry point for any *SPICE* program is the *netlist*. The netlist is a standard protocol for creating any *SPICE*-based circuit model. Even highly advanced *SPICE*-based models, such as Intusoft's (**www.intusoft.com**) *ICAP4*, which uses a drag-and-drop graphical entry pad, still generate a standardized *SPICE* netlist that can be read by any other *SPICE* program. This is great news, particularly if you collaborate on circuit design tasks.

Let's look at a simple netlist for a series resonant tank circuit, then go over it line by line. Below we have a circuit that is excited with a 100 V ac signal generator or voltage source, and has a 1 µH inductor, a 1 µF capacitor, and a 1 kΩ resistor, in that order.

```
series resonant circuit
V1 1 0 ac 100
L1 1 2 1e-6
C1 2 3 1e-12
R1 3 0 .01
```

The first line "series resonant circuit" is the netlist name, and most (if not all) *SPICE* programs require this on the first line. It's a good idea to name this something you will remember.

The second line is the voltage source, V1. The first number after the source name is the first node (terminal) of a component; the second number is the second node of the component. In this case, we have the voltage source between "node 1" and ground. Ground is always "node 0" and must *always* exist on any *SPICE* netlist. Interestingly, the order of the nodes for any individual component doesn't matter; that is, we could have listed this line as V1 0 1 ac 100. After the node numbers, we have the component values or parameters. In this case, it's an ac generator, with a voltage of 100. Simple enough.

The third line is the inductor. It is between node 1 and node 2. The value of the inductor can be listed in numerous ways; here we've used exponential notation, "1 times e to the –6 power" to designate the value in henrys (we could have used 0.000001, as well). Net lists, by the way, are a very forgiving of formatting, and you can slip in an extra space or line and it will nearly always figure out what you mean — it's very "Unix-like" in this regard.

The fourth line is the capacitor, which is between nodes 2 and 3. Its value is likewise expressed in exponential notation, $1e^{-12}$ F.

The fifth line is the resistor, between nodes 3 and 0 (ground), expressed in "normal" notation: 0.01 Ω. This is a good example of where you can have a "divide by zero" error during the analysis cycle if you don't have a finite value of resistance here. It can be as small as you like, but it can't be zero.

And there you have it, a text-based *SPICE* circuit wiring diagram, a virtual schematic that is understandable by any *SPICE* program.

ENTERING ANALYSIS

Once you have a netlist, your textual circuit model, you have to figure out what to actually do with it. This is the analysis part and it's also where there is a bit of variety between *SPICE* engines, although the nomenclature is still pretty standardized. In most modern *SPICE* programs, you don't see this intermediate nomenclature, as it is generally handled by mouse clicks. While you won't usually be editing the analysis files, you should know they're there.

In our case, we have a resonant circuit, so it might be nice to find out where its resonant frequency is. We can do this with a frequency sweep. Notice in your netlist that it doesn't say anything about the frequency of the signal generator. But if we do an ac analysis of the circuit, the analysis setup will ask a few things. How this is handled will depend on the "flavor" of your *SPICE*, but at the very least you will be asked for the beginning frequency, the ending frequency, and either the frequency increment or the number of points. You may also be asked whether you want the increments uniformly or linearly spaced, logarithmically spaced, or some other format.

Obviously, the resonant frequency of the circuit needs to fall within the frequency range you select. This is one of those cases where you need to know the answer before you give it to the computer to figure out. If you only "sort of" know the answer, it's a good idea to use the *logarithmic* increments option, which will let you sweep a huge range of frequencies with a limited number of data points. After running a frequency sweep in this manner, you should see something resembling a resonant peak. Then you can readjust your beginning and end points to include the established resonant frequency and re-run the frequency sweep, perhaps using a lot more data points.

When *SPICE* performs an analysis, it produces an array of numbers, each corresponding to a selected node in response to the independent variable; in this case, frequency points. The output values at each node can be voltage, current or phase, or combinations of all three. *SPICE* is very aware of KVL and KCL, and, in fact, performs most of its operations based on

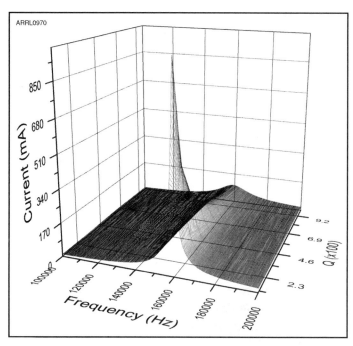

Figure 14.1 — Here we see the relative current through a series resonant circuit with varying values of Q. It's a great illustration of the value of multidimensional graphing.

Kirchhoff's Laws. Most modern *SPICE* programs also include the graphing software for plotting these arrays; this is how you get an immediate frequency response curve, for instance. For casual use, the internal plotting programs of most *SPICE* versions are more than adequate; however, it's sometimes interesting to look at the output arrays directly and twiddle the numbers according to one's requirements.

One particular trick I like to use is to plot the voltage response of a filter or other tuned circuit with a range of different Q values using a 3D surface or mesh plot. This method can be used when stepping any other circuit value as well, and besides giving you a really cool-looking graph it allows you to see at a glance the optimum response. The graph in **Figure 14.1** shows the amplitude response of our sample-tuned circuit with 10 different values of resistance, stepped between 0.01 Ω and 0.1 Ω in equal increments. Again, this is a snap to do with *Origin*.

LESSER OF TWO, OR MORE, EVILS

One of the more important functions of computer modeling is the *optimization* of solutions. In some delightful cases, a problem may have only one right answer, such as the resonant frequency of a simple series circuit in the above example. However, there are many cases when there may be several right answers to a problem. Just as frequently, there might be no right answer at all, in which case you may have to settle for an answer that's not *too* wrong. Finding the least repugnant solution of complex problems with many variables — some of which may have contradictory results — is the domain of computer optimization. It lets you choose the lesser of two evils, as it were.

Let's look again at our simple series resonant circuit to show how computer optimization works. For the component values given, as we've discussed, there is only one answer to the question: What is the resonant frequency? (It happens to be 159 kHz). On the other hand, if we were to ask what combination of L and C will give us 159 kHz, there are now an *infinite* number of possible solutions. For any arbitrary choice of L, we can find some value of C that will fulfill the requirement that $X_L = X_C$, the definition of resonance. In some cases, the choice of values for L or C may be dictated by what parts we happen to have lying around. On the other hand, we may have some other constraints to work within. If, for instance, we have a requirement that the value of Q for a circuit must be less than 100, we now have some limitations on the actual values of L and C that we can use. We know that Q = X/R, so if we have a fixed value of 0.01 Ω, again referring to our sample problem, our X_L is limited to values below 1 Ω. So the Q constraint ultimately limits the available values for L *and* C. Is this a good or bad thing? It's good in one regard: by limiting our choices, it makes the thinking part of our task easier.

However, what if we are able to alter the value of R1 at will? This again allows us an infinite combination of L and C that will fulfill the resonance requirement, and also allow us to choose Q at will. So, what *is* the optimum value of L, C and R, if we can now move any of them around at will, and yet still have the correct answer (actually *two* correct answers)?

In this case, there is no optimum solution, since we have an infinite number of equally good combinations of all three components. Fortunately (or unfortunately, depending on your point of view), we seldom have to deal with this problem.

Why is that? In this case, it's because we don't have control over a good part of the total R of the circuit; that is, the "parasitic" resistance of the coil. In reality, we will never have an infinite range of total circuit resistance to choose from. If the coil has 1 Ω of resistance, that limits our absolute lowest value of circuit R to 1. This will put an upper boundary on the value of Q that we can achieve.

Now here's an interesting little kink that shows how even the simplest *real* components can present us with an optimization problem.

THE Q CONTINUUM

We all know from our radio experience that a plain old inductor has a certain value of Q. This value is defined in the same way as for any complex circuit: $Q = X_L/R$. The R is the fixed value of wire resistance in the coil. For a simple solenoid coil, the inductance is a function of the diameter, length, pitch of the winding, etc. So we know that the coil's Q is limited by the resistance of the wire it's made of. However, we can generally increase the Q of any coil above its initial starting value by winding more turns. Why does this work? Because the inductance of a coil, all things being equal, goes up as a *square* of the number of turns, while the resistance goes up linearly (in proportion to the number of turns); that is, the resistance goes up in proportion to the length of the wire.

In addition, we also know that it's the inductive *reactance*, not the inductance itself, that determines the Q. If $X_L = 2\pi fL$, then we know that the reactance is a function of the frequency of the ac voltage applied to the coil as well. So, in theory, we can increase the Q of any coil by operating it at a higher frequency, all other things being equal. However, we also know about this little kink called skin effect, which amounts to an increase in wire resistance as you increase frequency. So this tells us that at some frequency, the Q will start heading *south* because of skin effect. This complex tug of war between competing parameters might suggest to the alert reader that there is an optimum frequency of operation of any coil, if a high value of Q is the object. And said alert reader would be correct.

It should be noted here, from a purely practical standpoint, that it's almost always easier to lower the Q of any circuit than it is to increase it. Just about anything we do has more losses than we intend, and this usually translates to lower Q than we would like to have. If Q happens to be too high, we can usually lower it by adding some resistance somewhere. This isn't always an option, however, as adding resistance might increase noise or some other unwanted side effect.

There is, by the way, an optimum form factor for a solenoid inductor; that is, it gives you the most inductance for a given length of wire, and hence the highest Q. Such a coil is just slightly fatter than it is long, a diameter-to-length ratio of about 8:7. The loading coil on GLA Systems' famous Texas Bugcatcher HF mobile antenna is based on this ratio (which is, by the way, an excellent case study in complex system optimization, done the *hard* way.

This leads us very nicely into the next chapter, which delves into variations on antenna modeling.

Antenna Modeling from the NEC Up

The first time I used a computer to actually *do* something was at HIPAS Observatory to study a phenomenon known as field aligned irregularities (FAIs), which are essentially "tubes" of plasma that surround magnetic field lines. I was tasked with building a UHF radar system to operate at 426 MHz, where there was an FCC allocation for experimental transmissions. I decided a large Yagi array, similar to an amateur moonbounce setup, would offer all the gain we could want. But we had some concern that cookie cutter moonbounce arrays had unacceptable levels of side lobe energy, as we needed a cleaner pattern to eliminate ambiguous reflections from the region of interest, namely the aurora. Fortunately, we had the famous Lawrence Livermore *Numerical Electromagnetics Code* (*NEC-2*) software at our disposal, which allowed us to model and optimize the antenna design for the minimum side lobe content consistent with good power gain. It fully *automated* that process, and what a difference it made.

NEC-2 was the original antenna modeling program from which all modern antenna modeling software derives. Originally written in Fortran, a few years later it was rewritten in BASIC (*MININEC*) with a few embellishments. Many Amateur Radio antenna modeling programs, such as *EZNEC*, are based on the *MININEC*. The excellent free software package *4nec2* is based on the original *NEC-2* code (which is, by the way, the modeling software I use the most). *NEC-2* and *MININEC* each have their strong points; *NEC-2* has a better-developed optimizing algorithm, while *MININEC* is a bit better at modeling real grounds and such. This isn't a bad thing, since you should never to get too attached to any particular piece of software — always use the *best* software for any particular job, at least if you can afford it.

We know that a computer is no substitute for experience and scientific understanding, but there are times when the output of a computer is an astonishing facsimile of reality. My experience with that FAI radar array was certainly a case in point.

FIRST STEPS AND PRIORITIES

With the right approach to antenna (or other) modeling, you can have a great deal of confidence in the answers you get from your computer. After doing this for a few decades, I very seldom encounter unpleasant surprises. The way to achieve this level of confidence (not only in your computer but in your use of the computer) is to start by modeling a lot of antennas that you're already familiar with.

Model a dipole and see if the pattern matches what you know a dipole is supposed to look like. (You do know what a dipole pattern is supposed to look like, don't you?) Next, try modeling a two-element phased array and see if you can get a perfect cardioid pattern. Hint: you need to drive the antenna currents 90° out of phase — not the voltages. Some modeling software provides actual current sources; the original *NEC-2* code does not. If you can't get a pattern you expect, you are quite likely not using current sources. This is probably one of the most common *NEC* modeling errors.

Any antenna modeling program that includes optimization will ask you what priority you place on any particular parameter. For instance, you can place top priority on forward gain and a lesser priority on front-to-back ratio, or some other factor. However, when optimizing any complex antenna, it is strongly recommended that you try to optimize only one parameter at a time. If you attempt to optimize two conflicting parameters, you may find that you never get an answer, that the program never converges. This is the entry point to optimization purgatory, which may do little besides test the heat tolerance of your computer's central processor.

As a cautionary note, it's always a good idea to look at the CPU usage of your computer when doing optimization programs. You might be surprised how *hot* these programs can make your computer, especially if stuck in a loop. I've destroyed a processor or two in my time doing this sort of thing. A good optimization program will usually recognize a setup that looks like it may not converge, and thus throw a processor meltdown-saving error, but don't count on it. If you don't get an answer in a minute or so, bail out of the process and adjust your "demands" slightly.

A LIGHT DISPLAY FOR MOM

NEC uses what is known as the *method of moments* (MoM) to provide the answers you seek. We know that the distribution of current along an antenna varies continuously, generally in a sinusoidal fashion, as you move away from the current source. The familiar "half-sine" distribution along a ½ λ dipole is a simple case in point. Obviously, since there are an infinite number of current values between any two points along the wire, strictly numerical solutions would be impossible. So the antenna is "chopped" into a finite number of segments, where each segment has equal current along its length. Thanks to some clever differential calculus, *NEC* is capable of interpolating with surprising accuracy the current distribution along the wire with remarkably few segments. And, if you can calculate the current distribution accurately, you can calculate the radiate field with equal precision. At first blush, *NEC* seems to work "a lot better than it should."

One simple and graphic demonstration I've developed to show the *NEC* segmentation process is to build a dipole antenna out of a string of miniature Christmas lights (**Figure 15.1**). This demonstration always elicits a chorus of "oohs" and "aahs" from a rapt audience. I feed a few watts of RF into the antenna from an HF transmitter, and the brightness of

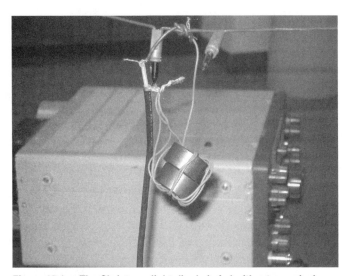

Figure 15.1 — The Christmas light dipole is fed with a transmission line balun, lashed together in classic research lab bailing wire methodology, proving that great science can be done under severe restrictions of funding and time, and other forms of duress. Sterling Muth, WL7TV, photo]

the bulbs follows the magnitude of the current along the string of lights. The brighter bulbs are near the feed point, and there are increasingly dim ones toward the ends. The light string very clearly illustrates a *standing wave*, even if there's only one bulb every six inches or so. If I double the frequency of the transmitter, two cycles are evident. If I run the third harmonic, there are three ½ λ.

Each lamp's brightness is proportional to the current in the middle of the segment in which it resides, even though we know the current along that segment is on a continuum — there's an infinite number of different values, even between each lamp.

By the way, in case you were wondering, this is *not* a recommended transmitting antenna for actual use, even during the festive holiday season. Not only is most of the power dissipated in the lamps, but the resistance of the string of lamps varies drastically with intensity. Needless to say, this would not make most transmitters "joyous."

LET'S RECIPROCATE

Central to all electromagnetic theory is the principle of *reciprocity*. We know that a charge carrier itself (such as an electron) is perfectly reciprocal. We accelerate an electron by applying an electromagnetic field; we create an electromagnetic field by accelerating an electron. Reciprocity allows us to use *NEC* antenna modeling with equal effectiveness for both transmitting and receiving antennas. However, we need to be careful to *interpret* reciprocity correctly and not make it "say" things it doesn't.

First, and most important, antenna reciprocity applies only to antennas in free space. As we've demonstrated in our plasma discussions, ionospheric propagation is not reciprocal. But even in the case of ground wave propagation, we have non-reciprocal factors. Ground absorption of radio waves is often highly non-linear. At MF, such as in the AM broadcast band, the near field losses can be dramatic, due to heating losses. At long distances, however, ground losses from an "incoming" low angle signal can be almost insignificant. The Beverage antenna relies on this real (but minuscule) ground loss for its operation. Ground losses are also highly polarization dependent.

Although antenna gain and pattern are indeed reciprocal, when it comes to transmitting and receiving we have some differences in other aspects of antennas. For instance, a transmitting antenna, assuming there is no resistive loss in the antenna itself, radiates 100% of the power provided to it, while a perfectly matched receiving antenna transfers half of its intercepted power to its load (presumably the receiver) and re-radiates half of its power. This, by the way, is not a violation of reciprocity; in fact it is consistent with the behavior of "black bodies" and other thermal-related and cosmological phenomena.

Because all "practical" antennas, at least at "normal" radio frequencies, radiate a spherical wavefront, not a planar one, we see a phenomenon known as *capture area*, which applies only to receiving antennas. As an antenna of a given size is removed from the source of spherical radiation, it absorbs, or rather "intercepts," a smaller proportion of the radiated signal. On the other hand, everything a transmitting antenna radiates is, well, *radiated*; that is, 100% of what it transmits goes somewhere. In the case of a distant receiving antenna, it absorbs only a minuscule percentage of the radiated power from an identical antenna.

HOW NEAR IS NEAR?

With perhaps a few odd exceptions, the end goal of any antenna design is to produce a certain result in the *far field*. Antenna performance is determined by what the radiation pattern is hundreds of wavelengths away from the antenna itself. For simple antennas, such as dipoles or single-element verticals, we really don't need to consider the *near field* (sometimes referred to as the *inductive field*) behavior. However, the near field properties of an antenna are crucial for designing multi-element arrays, such as Yagis and even phased arrays.

There is no sharp line of demarcation between where the near field ends and the far field begins. However, from a classical electromagnetic standpoint, the near field attenuates linearly with distance from the antenna, while the far field attenuates as the square of the distance (inverse square law). There is always an overlap between near and far fields, but the near field loses importance rather quickly as you move away from the antenna by a wavelength or two.

From a purely utilitarian standpoint, antennas or antenna elements can be considered to be within the near fields *of each other* if there is measureable *mutual impedance* between them. This may run counter to the conventional wisdom that you can define the near field region in terms of some percentage of wavelength separation. According to the mutual impedance criteria, many long-boom Yagis have directors that aren't even in the near field of their own driven elements. On the other hand, very long elements (antennas with large *apertures*) generally have larger "regions of influence" than shorter ones. In other words, there is more mutual coupling between longer antennas, so their effective near fields might be deemed larger than normal.

Mutual coupling is one of the reasons we must use current sources when modeling phased arrays. If there were no mutual coupling, each element of a phased array would have the same impedance as its free space impedance, and relative currents in each element would perfectly track the voltage applied to the element. However, mutual coupling not only changes the radiation resistance of each element, but also adds reactance, all of which make the use of voltage phasing invalid.

NEC-2 and *MININEC* handle near fields with aplomb — most of the time. One can sometimes encounter some odd results with very closely spaced elements, or with an element very close to a ground plane. Such conditions result in extremely low radiation resistances, which again can cause a lack of convergence. Or, more frequently, this condition can yield a result that is mathematically valid, but impossible to build physically. A case in point is if you model a very low dipole over an infinite, perfectly conducting ground. The gain of the dipole (straight up) will increase toward infinity as the height approaches zero! Unfortunately, this also results in a radiation resistance of zero, in which case the efficiency of any real antenna would also approach zero. (I suppose if one had a cryogenically cooled ground plane, one could get an infinite gain, zero-height NVIS antenna). Again, common sense reigns supreme.

A FEW ODDS AND ENDS

When modeling VHF and UHF antennas, it is critical to accurately model the conductor diameter accurately. This is seldom a significant factor when modeling HF wire antennas, where the length-to-diameter ratio is huge. When the diameter of a conductor is a significant percentage of the diameter, everything changes — often profoundly. As the number of elements increases, as in the case of long Yagis, the error caused by improperly scaling the diameters of the elements increases astronomically.

Here's a close corollary to the above: Don't fiddle too much with a good design! Folks have been building and modeling antennas for a long time, which is why some designs keep cropping up incessantly. Make changes in small increments; this applies whether modeling or actually building.

Always change just one variable at a time. This applies to every scientific discipline, but many hams get carried away by the convenience of modern antenna modeling programs. Every once in a while, by sheer luck, a decent antenna design can come about by means of an "optimize everything at once" approach. This won't happen to you. Remember my cautionary tale about purgatory.

Don't feel obligated to use a whole lot of segments when setting up a model. I have never needed more than 11 segments on any element. The only time you may need a lot more segments is if you have large diameter conductors or conductors with sudden bends or other transitions in them. Again, compare your models against known working antennas to get a feel for just how much detail you need. You do want to get in the habit of using an odd number of segments, however, especially if you use low segment counts. This allows you to place signal sources in the middle of a driven element. (You can safely ignore this rule if you mainly model verticals or other monopoles.)

The antenna is a deceptively simple device; in most respects it is little more than a piece of wire. And yet, that lowly wire is our only intermediary between the invisible world of electromagnetism and instrumentation. The more you understand about antennas, the more you will be able to understand the universe. Modern antenna modeling has vastly increased our level of both in recent years. Take advantage of it.

16 Large-Scale DAQ and Networking

Some of the most rewarding scientific discoveries are made by mining scientific data accumulated over years, or decades, and across many thousands of miles. History is replete with instances of great work being done, only to be set aside when interest or funding waned. There have also been times when we've had to discard collected data — irrelevant or relevant — for lack of "brain power" to analyze it or storage space to maintain it. But things have changed. With today's instant worldwide communications, including the Internet, and marvelous new ways of categorizing, analyzing and storing data, there's no need for this to ever happen again. Best of all, all these tools allow us to *decentralize science*.

There's no question that a lot of great science comes out of large facilities. Immense national laboratories, such as Lawrence Livermore, Los Alamos and Sandia, have cranked out continuous streams of scientific knowledge. We also look to the grand standard model-shaking discoveries from Big Science facilities like the Stanford Linear Accelerator (SLAC), Jet Propulsion Laboratories (JPL), and international facilities such as CERN. But they have their limitations, too. They're very expensive, often in far-flung areas, and can suffer from geopolitical constraints.

The great news is that, while "official" scientists tend to congregate in great facilities, scientific phenomena have no regard for geographic, political, or economic boundaries, and doing remote science is becoming the norm. Especially when it comes to Big Science, *remote sensing* is the name of the game, and this is where radio amateurs already have a leg up. Radio amateurs had an internet before the Internet: we called it packet radio. In fact, amateur packet radio was pretty much the main proving ground for technologies such as the TCP/IP protocol, upon which the World Wide Web is based. It's easy to forget that the Web only came to the public consciousness around 1995; it's even easier to forget that only radio amateurs were using anything resembling internet technology on a significant scale prior to that.

DECENTRALIZING SCIENCE

Because scientific phenomena do not recognize man-made boundaries, the concept of *distributed experiments* is extremely attractive, especially to those of us who don't live near major science hubs. It also offers vast potential to add significantly to the body of scientific knowledge. Let's look at an example from astronomy to get our creative juices flowing.

The world has only a handful of very large telescopes, such as the ones at Palomar Observatory in Southern California or W.M. Keck Observatory in Hawaii. These are immensely expensive and sensitive instruments, and for many studies there is no substitute for the power of a really big telescope. On the other hand, there are literally tens of thousands of pretty darn good telescopes in the hands of amateur astronomers and radio amateurs (two largely overlapping groups). There are some major advantages to a lot of pretty doggone good telescopes over one really good telescope.

Big telescopes can look only at one place at a time. As a general rule, the more powerful a telescope is, the more restricted its field of view. Although there are some "all-sky telescopes" (very large telescopes that can scan vast areas of sky very quickly), they are even *more* than immensely expensive. But armies of amateur astronomers with modest telescopes can observe a lot of the sky at any given time, and are, at least statistically, more likely to spot an interesting or fleeting event. If two amateur astronomers, or a whole *lot* of amateur astronomers, should happen to see the same event, there can be immediate confirmation, thanks to immediate communication.

In the past, it could take years, decades, or even centuries, to confirm (or dismiss) the existence of some rare event. If 100 amateur astronomers around the world spot a large asteroid colliding with Jupiter, it's probably a fair conclusion that that the asteroid collided with Jupiter. In fact, the event can probably be confirmed, logged, and published in the time it takes to swivel a Palomar-sized telescope around to locate the event.

With a little more technical know-how, a very large telescope can actually be synthesized with a lot of smaller telescopes. The light-gathering capacity of any lens is a function of its area (aperture), but that aperture doesn't all have to be in the same place. Extremely dim objects in deep space can be made brighter by combining the images from a lot of smaller aperture telescopes. Again, with the proper communications networking, this sort of thing can be done in real time. At a greater level of sophistication, through the use of *adaptive optics,* one can actually achieve optical *phase coherence* between widely separated telescopes. Adaptive optics is just becoming available to the amateur astronomer, but it probably won't be too long before this technology is readily available at an affordable price.

REMOTE SENSING UP CLOSE

Closely associated with distributed experimentation is remote sensing. Remote sensing broadly encompasses any technology or hardware that gives you information from some location that's not readily accessible. It generally excludes any device that might be in physical contact with the object or phenomenon you're measuring. For example, hanging a radio transmitter around a buzzard's neck doesn't strictly fall into the definition of remote sensing, no matter how remote the buzzard might be.

Examples of remote sensing include radar, sonar, lidar, and active seismology. They all basically use the principle of "shooting a gun into a dark room and listening for whatever screams." One interesting technology that falls into the category is, ironically, X-ray crystallography, which examines the internal structure of crystals. In this case, the "remote" location is the center of an atomic matrix, which may be only microns in the "distance"; nevertheless, it is a region of the universe inaccessible by other means. The latest advance in X-ray crystallography is the X-ray *laser* or *free electron laser*, which is capable of such short pulses that it can be used as a sort of strobe light to observe the creation or destruction of chemical bonds *in real time*. Amazing!

But right within the home shack, every radio amateur who transmits a signal that is received by someone else is playing a role in remote sensing. In a large percentage of cases, the thing we're sensing is the ionosphere. We infer a great deal of information about what's way up there by how it affects what we see down here. All remote sensing methods have a certain intriguing degree of uncertainty. We have to make certain assumptions about the nature of what we're observing, so we start with a model that we hope is a fairly good one, and use our measurements to give us more detailed information about what we think we already know. Our confidence level, however, increases dramatically if we can construct a model using two *very* different methods. Here, seismology provides a fascinating case in point.

A good deal of what we know about the geology of the Earth is based on seismic sounding. We've only been able to actually drill a few miles or so to obtain actual *in situ* observations of what lies below us. Deep hole drilling barely scratches the surface of the wonders that lie beneath our feet. Our "three layer" model of the Earth (core of primarily nickel and iron, mantle, and crust) was primarily developed using seismic methods, looking at acoustic signatures generated by natural or, in later years, man-made disturbances. Modern seismology uses explosive charges to "ping" the Earth to obtain better-defined results than can be gleaned by simply relying on natural mini-earthquakes. More recently, ultra-wideband ground penetrating radar as well as VLF, ELF and ULF radio sounding methods (often down in the microhertz frequency range), has corroborated the three-layer model. Because these radio methods, which basically act as a giant metal detector, are so different from seismic sounding, they provide a pretty unbiased second opinion. We can always use a third or a fourth opinion, too.

AMATEUR RADIO'S LEADING ROLE

A big difference between Big Science and amateur science is that most of the "official" participants in the former do it as a full-time job. Radio amateurs who do radio science, for the most part, do it in their free time. This is why it's called *amateur*, which means the work is done without pay, *not* that it's done without expertise.

Because of their own full-time jobs, most Amateur Radio scientists couldn't babysit a telescope, ELF receiver, or a homebrew particle detector 24/7, even if they wanted to. Therefore, we have to find creative approaches to collecting data if we want to do any real meaningful collaborative science. The Search for Extraterrestrial Intelligence (SETI) Institute (**www.seti.org**), which formed in the mid-1980s, offered probably the first model we had for doing amateur science in a big way. Although I have some qualms about their scientific premises, SETI's methodology was certainly logical and admirable. It basically involved using "spare" computer processing power of dozens of participants' computers, an approach known as *distributed computing*. The project started when computer hardware was still relatively expensive, especially in memory devices. Today the average laptop probably has more number crunching power than the entire original SETI setup, but the principles can still be used to great advantage.

One of the achievements of SETI has been *serendipitous science*; that is, there have been a number of astronomical discoveries made that, while having nothing to do with extraterrestrial intelligence, are of great interest to general cosmology. Its Project Argus (named for a mythical giant with 100 eyes) is a technical initiative with the goal of assembling several hundred amateur radiotelescopes (using low-cost converted TV satellite dishes) in an "all-sky" network. There are about 150 installations

Saving Time for Skymapping

Among the most important recent advances for distributed science is the availability of accurate time references. To do good distributed science, you have to know not only "*where* you are" but also "*when* you are." The nearly ubiquitous GPS receiver allows us to know both. If you have a widely distributed sensor array looking at some time-sensitive phenomenon, you can *reconstruct* the item of interest if it can match up the *time stamps* of all the data points. This doesn't have to be done in real time. If the times of the relevant samples are known, they can be correlated later when the number crunching is performed. This is great news if you have limited bandwidth or limited availability communications channels between you (or other designated central point) and remote the data collectors. The remote data can be sent "home" whenever it finds an available path, and this doesn't need to have any relationship to the time of DAQ itself. Of course, this doesn't work for *every* type of distributed science, but it works in a lot of cases.

At HIPAS Observatory, in the mid 1990s, we exercised this principle a bit. One of our many projects was to create a *skymap*, a "topological" map of the ionosphere over interior Alaska. We knew there were a lot of bumps and wrinkles in the ionosphere, but we didn't know how deep and wide they were. So the idea was to employ "HF holography," essentially the same methods used for optical holograms, to create our map. We assembled an array of phase coherent receivers located around Fairbanks in as many accessible points as had an accurate time reference available. We then used one of the main "heater" transmitters at the observatory to launch a broad-beamed NVIS signal into the ionosphere. This signal would be reflected from a large area of sky, and received by each of the coherent receivers, both in phase and amplitude. We could thereby measure the time-of-flight of the transmitted signal to each receiver, using accurate phase measurement, and thus determine the height of a relatively small region of the sky.

We didn't have any communications paths back to the main laboratory, so we had to store the data from each receiver locally, using precise time stamps. The real trick here was that we didn't have GPS receivers available, either to generate the receiver phase reference *or* the time stamps. However, what we *did* have at each site was T1 carrier signal availability.

Because I had previously worked for our local phone company, we were able to place our remote locations at sites I knew had T1 references available (with the permission of the T1 customers, of course). A T1 carrier has an atomic standard time reference, which was as good as any GPS receiver, so we used that to lock in our local oscillator synthesizers (after converting the T1 Frequency of 1.544 MHz to something that made the synthesizers happy).

After a few weeks of experimentation, we checked the sites and downloaded the data from the respective hard drives, each of which had data recorded with National Instruments' (**www.ni.com**) *LABVIEW* software. We loaded the data from the various sites into a monster spreadsheet, lined up all the time stamps and plotted the data on a 3D *MATLAB* program one of our technical gurus had written. We obtained a surprisingly successful and revealing series of skymaps, each taken at about 1 minute intervals. Although we ran out of funding to pursue this much beyond the proof-of-principle stage, prove it we did.

as of this writing. This "multi-eyed" undertaking exploits the greatest strength of distributed science: its huge potential for capturing interesting data, even that which is not specifically being sought. Centralized Big Science, on the other hand, can't afford to follow rabbit trails, no matter how fascinating or fruitful those journeys might prove.

Hams were in the vanguard of distributed science prior to SETI, however. When packet radio first became popular with radio amateurs, it was primarily used for small text files, bulletin boards, and what later evolved into e-mail. Packet really got its start in earnest on 2 meters, but eventually spread up to the microwave bands as well as down to HF radio, although not terribly successfully without some modifications.

By today's broadband Internet standards, packet radio is abysmally slow and has largely dropped out of use on the amateur bands, except in the form of automated packet reporting system (APRS). APRS has brought some renewed interest in packet radio, however, and the principles used for APRS are particularly interesting and attractive for doing distributed science. While a lot of science requires vast number crunching capacity, as well as big fat communications pipelines, a lot of it does *not*, and a significant amount of DAQ is little more than data logging (meaning data acquisition with intervals on the order of seconds or minutes, not microseconds or nanoseconds). Old-school packet radio is more than up to the task of relaying this kind of data back to a larger hub where it can be processed.

But why would we use packet or APRS for doing science when the much faster Internet exists? Because, as we explained earlier, scientific phenomena are not subject to economic or geopolitical whims, unlike the Internet. Science happens where it happens, and often it happens where there is no Internet, say in some remote parts of interior Alaska, or Antarctica, or the ocean, or beyond Mars. Frequently, amateur packet radio is the ideal communications medium for doing science (see "Saving Time for Skymapping" sidebar). And, though the bandwidth is limited, at least on VHF, there is a lot more bandwidth available on the amateur microwave bands.

American radio amateurs have 12 microwave bands to use pretty much as we please. Not only are they excellent for high-speed data transmission, a nice thing to have for doing radio science, they are an incredible scientific resources in themselves. They are precious allocations we need to both study and safeguard. Thus endeth the diatribe.

Large scale data acquisition is just one more tool to allow us to see the "whole elephant," and like so many other scientific methods available to hams, is a work in progress.

Graphs and Graphics for a Big Picture

Some physics principles are just incredibly difficult to visualize until you see them, which is why scientists are so fond of graphing. Looking at data itself, no matter how well organized, can appear to be nothing more than random events, no matter how many eyes pore over the information. Graphs and diagrams, however, reveal trends and patterns that might otherwise be easily missed. To see the big picture we usually need to catch unusual, or even paradigm-changing, phenomena.

Making endless graphs of pointless data may have seemed like punishment during your formative years, but with modern technology, it's actually a whole lot of fun. It takes some discipline — perhaps even learning a *new* discipline — to properly utilize the new math tools at our disposal. But once you learn the ropes, you'll wonder how you ever got by without their help.

Back when we all did science by hand, we seldom had to deal with vast quantities of numerical data. When I was chief engineer of KJNP, a 50,000 W AM broadcast station in North Pole, Alaska, I was tasked with taking weekly field strength measurements of our two-tower directional antenna array. This involved traveling to a number of remote monitoring locations and setting up a tripod with a field strength meter on top of it, waiting for the DJ to pause so I could measure the unmodulated "dead air" carrier wave field strength. I'd return to my office and plot the points on a piece of polar graph paper to see if the pattern had changed significantly over the week. My yearly proof-of-performance measurements involved pretty much the same process, only a whole lot more of it! Still, I could do it with a pencil and a piece of graph paper.

Things have changed since then. Not only do we obtain much more accurate scientific data in nearly every scientific discipline now, we also have a lot more of it to deal with. Modern DAQ hardware and software collects data on the order of thousands, millions, or billions of points per second. Compare that with my antenna measurements, that might have represented only one data point every few minutes.

Fortunately, the ability to handle vast quantities of data has kept pace with the ability to acquire the data in the first place. When I was in engineering school, linear algebra was presented to us as not much more than a mathematical curiosity; the concept that it might actually be useful for us electrical engineers wasn't even suggested. All this has changed dramatically. All computerized scientific data manipulation takes great advantage of linear algebra, primarily in the form of multidimensional matrices and arrays, both of which are simply logical and convenient arrangements of huge collections of numbers. All modern scientific software, such as *MATLAB*, *Mathcad*, *Scilab*, *Octave*, *Origin*, and even the graphing functions of *Excel*, perform all the gnarly linear algebraic functions invisibly, letting you concentrate on the really fun stuff.

SEEING IN 3D

Some people have a difficult time grasping the concept of multiple dimensions, at least beyond the fairly tangible spatial three dimensions we usually experience. Although a lot of what we deal with in radio science involves normal spatial dimensions, any physical quantity or parameter can be expressed as a separate dimension. This is particularly useful when we have multiple parameters that are related to one another. For instance, let's take a sample of a distant radio signal at some point in space. Let's call this Sample Point A. At this single point, which is itself defined by the normal three dimensions, we can *also* have signal strength, frequency, polarization, and direction of arrival. So Sample Point A has actually seven dimensions, as follows:

- X coordinate
- Y coordinate
- Z coordinate
- Strength
- Frequency
- Polarization
- Direction of arrival

Each of the seven dimensions listed above is associated or correlated with Point A. So, how can we express all these dimensions on a single graph? Well, the location on the graph (with 3D rendering) is simple: we locate it at position XYZ. We can, furthermore, express the signal strength at that position by means of color, say with "bigger" signals shown as "hotter" colors and weak signals shown as "cooler" colors (a very common practice). Frequency could be expressed by the width or size of the plotted line or point. The line style could then be used to display the polarization; for example, a solid line could represent clockwise circular polarization, while a dotted line could express counterclockwise circular polarization. All the scientific software packages mentioned above are capable of rendering this sort of thing.

Now the last dimension, the direction of arrival, could be a bit tricky to plot on our graph, since this dimension itself has three dimensions: X, Y, and Z. However, this is a great example to introduce the next important concept: arrays of arrays. That leads us into arrays of arrays of arrays. And *that* leads us into, well, I think you get my drift. Let's look at a large VHF antenna system as an example of multi-level arrays and how we might do some interesting matrix math on such a construct.

As I mentioned earlier, at HIPAS Observatory we built a large UHF Yagi antenna array (426 MHz) for an ionospheric radar, specifically to investigate FAI phenomena (the "wrinkles" one normally sees in a visible aurora and that clearly show up on UHF radar as well). The composite antenna array consisted of 16 phased Yagi antennas in a 4 × 4 pattern (four horizontal, four vertical), each Yagi having 17 elements (or 17 elements deep). We put the whole thing on a 20 foot tower, with both an elevation and an azimuth rotator.

Now, how would one describe this antenna in useful mathematical array notation? It would be expressed as a 4×4×17 array. Since each Yagi itself was an array, the composite structure was an array of arrays; or more precisely, it was a 3D array of 2D arrays (each Yagi being a 2D array). With enough time, money, and patience, we could have built six of these steerable arrays, and placed them all in a row, to create an array of arrays of arrays. Then perhaps a few miles away, we could have built another six-tower array, ending up with an array of arrays of arrays of arrays.

Needless to say, this sort of thing can be a bit difficult to keep straight in your head, unless you happen to be naturally good at multidimensional thinking. Now despite this somewhat simplistic and mechanical description of multilevel arrays, the manipulation of multilevel data arrays is a common scientific discipline. It's also at the core of all your modern scientific graphing and most modeling software. In fact you probably do a certain amount of array manipulation without even thinking about it, such as when you sort your email addresses, or even balance a checkbook.

Let's ease into array manipulation with some familiar and intuitive examples of arrays and see how we can get some fascinating and impressive graphics.

MAKING SENSE OF SENSOR ARRAYS

In our description of that multi-tiered Yagi array at HIPAS, we didn't look too closely at the internals; we treated the "array of arrays of arrays of arrays" as pretty much a functional block — in this case, one big huge antenna.

In Chapter 9 we saw how even monster antennas such as this really behave as a point source, once you get a few dozen wavelengths away from the thing. In general, large antenna arrays exist in order to create stronger radio signals than would ordinarily be possible, or to receive weaker signals than would ordinarily be available, but there's still nothing particularly special in the way they radiate.

Sometimes building a very large array of instruments serves a very different purpose than simply making it a more sensitive or powerful instrument. Take the case of particle detection, the search for new or even familiar, subatomic particles. While tracking single particles of any kind can be a bit tricky, it's well within the means of any radio amateur to detect, say, a stream of electrons. It takes only a rather simple current transformer of a toroidal form to enclose an electron beam within its core. Current transformers can be purchased or even homebrewed, but with a bit less accuracy.

Such a transformer allows you to determine the amount of current flow in an electron beam, and its direction — well, almost. Any electron flow going through the hole in the transformer in one direction will read "positive," and any flow going the other direction will read "negative." At any point in time the polarity of an electrical current in any two-terminal device, such as a current transformer secondary, can only be one of two values: positive or negative. What if the electron beam is passing through the hole at a 45° angle? Or 1°? Or at any angle that's not perpendicular to the core? The best you can really do is to determine the sense of the direction of the beam; that is, the sign of the direction. While this is very crucial information, don't despair. Just hold that thought for a bit.

Now let's take a slight detour and talk about direction finding antennas. As you might suspect, determining the direction of an electron beam is not the same as finding the direction of a radio wave, but there are similarities. One of the most familiar and effective direction-finding antennas is the small loop. It has very sharp nulls perpendicular to the plane of the loop, which are very good for pinpointing a distant station — except for the fact that you've got *two* sharp nulls, 180° apart. What is missing in this antenna is a *sense* indication. We, therefore, have a serious *ambiguity* problem; that is, there are two possible answers, only one of which is correct. One traditional way of resolving this ambiguity is by means of a sense antenna, which actually modifies the pattern of the loop into a unidirectional pattern. But it's really using two antennas to achieve this: a small vertical whip and the main loop antenna.

As it turns out, nearly any physical quantity we want to measure in radio work will have some ambiguity function associated with it. Direction of arrival, phase or polarization are all factors that can be "off" by 180°. One of the beauties of using complex numbers to express such values is that the ambiguity is easily resolved. Most physical quantities in the universe can actually be expressed as complex numbers. The universe positively seems to like complex numbers; we have countless physical systems that exist along orthogonal or perpendicular axes. In electromagnetism, we have the electric and magnetic field at right angles in the spatial domain. In a capacitor, we have current and voltage at right angles in the time domain. In an inductor, we have this 90° time domain phase shift. In a mechanical system, such as a satellite in orbit around a planet, we have acceleration at right angles to the velocity. When this occurs, there is no energy gained or lost; it's another example of an orthogonal system.

Now, returning to our electron beam detector, we see we started out with a "sense" detector, but no actual angle detector. A loop antenna is a very good angle detector, but not much of a sense detector. What we want is an *array* of detectors or sensors. One of the prime motivations of building large arrays of detectors — whether it's for looking at particles, gamma rays, or plain old radio waves — is to eliminate ambiguities.

So how would we go about modifying our detector to really tell us which direction the electron beam is going? Well, we could place another current transformer a foot or so behind and parallel to the first one, positioned so that only a beam passing nearly perpendicularly though the first transformer will pass through the second one. Then we could know with a fair degree of confidence the direction of the beam travel, *provided* the beam is traveling in pretty much the direction we expect it to travel in the first place. *A priori* (Latin for "from the earlier," and here meaning "already formed") knowledge is actually one of the best ambiguity-fixing tools there is. We usually don't have random electron beams squirting around the laboratory, or in free space. Electron beams are generally formed in a decidedly deliberate fashion, say, in a vacuum, with a device *known* to squirt electrons in a certain direction. But this example simply shows that an array of two current transformers is better than a single current transformer.

In reality, it's a bit more practical to detect random charged particles by means of *capacitor* plates, since charged particles carry, well, a charge. If you look at a photo of the inside of an atom-smasher's detection chamber, you will likely find vast arrays of capacitor plates or pairs of plates surrounding the target or reaction area. These plates will be generally grouped in orthogonal "vertical" and "horizontal" pairs (we have to use vertical and horizontal advisedly here, since these orientations may have no connection with normal terrestrial reality). As charged particles pass between these plates, they create an electrical charge offset between the plates, unless they happen to pass right through the center, in which case the voltages are perfectly balanced out. It sort of acts like an oscilloscope in reverse: Rather than using deflection plates to deflect electrons, the plates "see" charged particles passing between them. So we know that if a particle passes through the plates "off-center," in either the horizontal or the vertical plane, or at any angle in between, we can detect the momentary position of the particle. A "dead center" particle will leave no detected voltage — but so will no particle at all! We seem to have stumbled across another interesting type of ambiguity, haven't we?

One way to resolve this ambiguity is to set up another array of *unbalanced* capacitor plates that simply detect the presence of a charge without telling us much about its position or direction of travel. Of course, we have to coordinate the response of these particle-presence plates with the actual particle position plates. In addition, we have to make sure the existence of this unbalanced plate array doesn't upset the balance of the *balanced* plates, which may number in the hundreds or thousands (and which is also the sort of thing that makes building Big Science experiments so much fun).

Note that this particle detector is an extremely simplified device, set forth only as an example. Most of the interesting particles scientists are looking for these days don't follow straight trajectories, but rather spinning, curling, looping, and rapidly accelerating paths, all acted upon by vast numbers of known and unknown forces and fields. If you took a typical high school physics class, you probably built a simple cloud chamber that traced the paths of various particles in a fascinating, direct, visual manner. It is these convoluted paths, among other factors, that identify the particles in question. This is why mathematical array manipulation is so central to this kind of science. By using extremely fast DAQ in concert with large detector arrays, scientists can reconstruct the history, in both time and space, of the fleeting particle and with a good degree of confidence.

FOCUS ON SYNTHETIC APERTURE

With a nod back to our "whole elephant" analogy, we now see that to get the big picture of some Big Science phenomena, it helps to have a lot of sensors. This isn't always convenient, but the good news is that in many cases we don't have to look at the whole elephant at the same time. Nothing demonstrates this better than the *synthetic aperture radar* (SAR).

Scientists often wish they had a big antenna — I mean a *really* big antenna, say, thousands of miles across. We know from basic electromagnetic theory that the bigger the antenna, the more gain (or sensitivity) it has, at least potentially. This applies from the lowest radio frequencies up to optics and above. Just as bigger telescopes are better than smaller ones, bigger antennas are better than smaller ones. Luckily, we can make a small antenna look like a much larger antenna by moving it along a path that would follow the contour of the larger antenna, if it actually existed.

Let's imagine that, as inquisitive radio amateurs, we are suspicious that life may exist on a planet in orbit around Betelgeuse, just to pick a star. If it does exist, said life might be trying to contact us on 80 meters. Imagine also that any self-respecting space aliens know Morse code and are using it in their attempts to reach us, even if it's really *slow* Morse code, say one character a month. It's probably a fair assumption that under Betelgeusian Amateur Radio regulations, any signal we'd receive from there would be somewhat on the weak side.

After doing some back-of-the-envelope calculations we determine that a collinear array of 80 meter dipoles stretching from Los Angeles to Newington, Connecticut (about 2,887 miles), might give us the sort of gain we need to receive a Betelgeusian 80 meter CW signal. This will take 115,480 half-wave dipoles, give or take a few, all combined precisely in phase. Needless to say, aside from the expense of the 115,480 runs of coax cable and the signal combiners necessary to complete the project, it might be difficult to get approval from all the homeowners' associations along the route, unless you build the antenna in the middle of I-80. There must be an easier way. Indeed, there is.

We can put just one 80 meter dipole on a truck hauling a couple of tandem 60-foot trailers behind it, then add an 80 meter receiver and DAQ system in the cab, and drive the thing across the country. We can then simply take a sampling of the signal with our high-speed DAQ system every $½ \lambda$ along the journey. When we get to Newington, we combine each of the 115,480 RF samples, correcting the phase of each sample for the change in time of the dipole intercepting the wavefront at each point, and *voila*: a space alien QSO! We just *synthesized* a 2,887 mile long antenna with only one common, ordinary dipole. It just took a little longer than normal.

It probably goes without saying that this method doesn't work too well with moving targets, and we have to assume Betelgeuse is in the same place it was at the beginning of the trip. And it wouldn't work if the Betelgeusian ham was sending high-speed CW or voice. It's also extremely important to note that the phase correction mentioned above is no mere footnote: it is the key to the entire concept. Without the phase correction, we have no gain whatsoever over a single dipole being dragged across the country.

Finally, we should emphasize that no brilliant concept is anywhere near as easy to implement as it is to conjure up. In reality, to complete our imaginary QSO, the speed of the truck would have to be extremely precise and constant, nearly as precise as the speed of the radio signal itself. Any departure from this standard would result in jitter, which translates to noise after all the number crunching is done, which severely degrades the viability of the concept. For Earth orbiting satellites, however, synthetic aperture works just as advertised, or nearly so.

MORE LOWDOWN ON ELF

SAR methods aren't only good for observing very small or far away objects, but also when we are dealing with ELF. We've already touched on a number of challenges encountered when we attempt to create ELF radio waves, at least to do so with any degree of efficiency. To be a reasonably efficient radiator, an antenna needs to be on the order of at least $\frac{1}{10} \lambda$, but it's much better if it's $\frac{1}{2} \lambda$ or longer. Receiving antennas can be much smaller, where "efficiency" can have very different connotations, but we still run into some serious limitations when working with wavelengths on the order of hundreds or thousands of kilometers.

Although not a synthetic aperture array in the strictest sense, we can synthesize a very large antenna by using just a few very widely spaced array elements, even if these elements are stationary. This is sometimes known as a *very sparse array,* and while not quite as effective as a huge array with all its parts present, the sparse array can be a very useful scientific instrument for ELF. The nice thing about a very sparse array is that the elements don't even have to be uniformly distributed across the actual "antenna." One fascinating application of this concept is NASA's Radio JOVE project (**http://radiojove.gsfc.nasa.gov**), which fits in very nicely with the entire tenor of this book. This long-term experiment consists of lots of low-frequency radio receivers installed by high school and middle school students across the entire nation, each looking for radio signals from Jupiter or the Sun, and potentially from some even more distant astronomical objects in the future. Although the low frequencies associated with astronomical phenomena present serious challenges in terms of sensitivity, it is actually much easier to obtain phase coherence between receivers at ELF/ULF and below. And, as with our previous example of SAR, in project JOVE all the number crunching can be done after the fact. As long as the precise location of each receiver is known, one can reconstruct the distant wavefront in a relatively simple manner.

In the remaining pages we look at physical apparatus and graphical methods to help you achieve a coherent view of modern scientific methods.

Rolling Your Own — Building Instruments for Radio Science

A good deal of radio science is uncharted territory, a vast intellectual field just waiting to be explored. But one of the drawbacks (and benefits) of being first in a field is that you have to bring your own tools along. For any scientific investigation, there will be a few tools you'll need that you can't pick up at the local hardware store and you have to be innovative and resourceful. Luckily, in our exploration of radio science much of the hardware necessary can be found in your own radio shack, but if not, you've got the skills to build it.

Every radio amateur should build at least one crystal radio receiver (commonly just called "a crystal radio"). I've built numerous such devices, and they never fail to intrigue me. They represent the simplest interface between the hard, physical world of wires and the abstract, intangible world of radio waves. One of the best resources for parts and information on simple, and not so simple, crystal radios is the Xtal Set Society (**www.midnightscience.com**).

The crystal radio is one of the best devices available, at any price, to demonstrate such principles as reactance, resonance, basic antenna principles, energy storage and transfer, plus a whole lot of other subtle scientific ideas. But sooner or later, however, if you want to do real radio science, you're going to need a more sophisticated receiving apparatus. You'll have to deal with some extremely weak signals and be able to separate them from various sources of noise (although in some cases, it will be the noise *itself* that's the object of interest). You're going to have to stretch the limits of your understanding and application of radio reception techniques.

Every kind of radio receiver familiar to the amateur operator is used in radio science, including tuned radio frequency (TRF) receivers and superheterodyne (superhet) receivers, both the single conversion and multiple conversion variety. One of the most important receivers for scientific study, however, is the direct conversion type, in particular the so-called "lock-in amplifier," which is really nothing more than a specialized direct conversion receiver.

THE DC RECEIVER GROWS UP

The purpose of a radio receiver is to receive radio signals. That part is fairly self-evident. Early radio receivers, such as the aforementioned crystal sets, were relatively "deaf." One could improve the sensitivity of a crystal radio by adding amplifiers, either before or after the crystal. The problem with indiscriminately adding amplifiers to a radio receiver is that they amplify indiscriminately; not only will they amplify a desired weak signal, but they will amplify noise and other undesirable radio stations as well.

The development of the TRF amplifier-based receiver addressed this problem to some degree. Many high-quality AM broadcast receivers used multiple tuned circuits before the detector stage, which added greatly to the selectivity of the radio. With a well-designed TRF set, one could easily receive distant stations located very close (frequency-wise) to other nearby stations. However, a fundamental characteristic of tuned circuits meant that the TRF receiver's selectivity varied widely from one end of the "radio dial" to the other. The top end of the AM broadcast band is about three times the frequency of the bottom end, and it was very difficult to build a radio with consistent performance across the entire range.

Enter the superhet. This design uses a frequency converter to shift the desired incoming radio frequency to an intermediate frequency (IF), which was really just a fixed-tuned TRF receiver. The performance of the IF and everything following could be optimized for sensitivity and selectivity, and all was well in the world. Well, almost, but not quite. Another immutable law of physics reared its head.

Any conceivable frequency converter generates what is known as an image frequency in addition to its desired output frequency. A frequency converter consists of a mixer stage, which is fed two different frequencies, and so generates both the sum and the difference between the two input frequencies. In the case of the practical superhet receiver, a local oscillator (LO) is fed into the mixer along with the desired radio frequency. The IF stages of such a receiver will, therefore, respond to a radio signal both above and below the LO frequency by the same amount. Manufacturers mitigated the image problem by selecting an IF that would have the image frequency fall outside any occupied radio frequency band. For a long time, 455 kHz was the IF of choice. Any image frequencies were above about 1.5 MHz, where there were no radio stations — until people started using shortwave radio frequencies, that is.

So image frequencies became a bit of a problem themselves. This quandary was partially solved by increasing the image frequency to a much higher value. In so doing, the image frequency was much farther removed from the desired frequency, allowing the natural selectivity of the front end to do the image rejection. But this created yet another problem: the overall selectivity was reduced, because the Q of the higher frequency tuned circuits was reduced.

As one might suspect, some brilliant minds came up with yet another fix: the dual conversion superheterodyne receiver. This design used a high IF and low IF, each stage requiring its own frequency converter and associated IF "strip." The high IF took care of the image problem, and the low IF took care of the selectivity problem.

Once again, all was right in the world for a while. The problem, as we saw, is that no mixer is perfect. Each one generates image products, but it's actually worse than that. A mixer not only generates sums and differences, but also generates sums of differences, and differences of sums, and sums and differences of harmonics (integral multiples) of not only the input frequencies, but the output frequencies as well. In short, a single mixer generates a hodgepodge of frequency products. These products show up as a series of "birdies" that even the best of commercial-grade receivers display. Most of the time, at least in Amateur Radio application, you can ignore or evade such birdies, but not with scientific radio receivers; you need to know that every signal you might receive is *real* and not a product or artifact of the receiver itself. Despite their shortcomings, dual conversion (and even multiple conversion) receivers are prevalent in amateur, commercial, and military communications and probably will be for a long time.

However, for scientific work, where ambiguous signals cannot be tolerated, a *very* old method of reception has become very popular once again: the direct conversion, or homodyne receiver. The direct conversion receiver is capable of the cleanest performance of any known receiver, and it is the receiver of choice for almost all scientific work.

Physically, the direct conversion receiver is very similar to the superhet, except that the LO is exactly on the frequency of interest. Selectivity is achieved not by means of tuned amplifiers, but by low-pass filters and amplifiers. These amplifiers generally work at audio frequencies or lower, and are very easy to design. Image frequencies do not exist because there is no difference (or sum) between the incoming frequency and the LO. It is an extremely simple and elegant solution to what had seemed an intractable problem in the radio science world.

The direct conversion receiver might have become the "standard" for just about any radio service 80 years ago, except for one drawback: it requires an extremely stable and clean LO. Modern frequency synthesis methods now make the direct conversion receiver a very practical option, which did not exist a few decades ago. "Free running" LOs of times past were unsuitable for effective direct conversion receiver design.

IMPROVING YOUR IQ

Let's camp out for a while on the venerable lock-in amplifier, a specialized version of the direct conversion receiver. This is one of the oldest and most effective tools for looking at extremely weak signals far below the noise floor. Traditionally, the lock-in amplifier is used for either ELF or even dc signals. One interesting application is for measuring the speed of heat flow through various materials by measuring extremely weak variations in resistance at various points in the sample.

The key to the lock-in amplifier's magic is a phenomenon called autocorrelation, a rather counterintuitive phenomenon, but that seems to be built into all of nature and is observed quite a lot in everyday activity. Our brains are the best autocorrelation devices on the planet; they are hard wired to see and hear patterns in the noisiest circumstances. The ability to recognize one person's voice in a crowded room and filter out all the rest is one common example. Scientists and engineers have been trying to match the human brain's pattern recognition prowess for decades, but they're still a long way from approximating what lies between our ears. However, the invention of the lock-in amplifier was a major step in this direction, at least for one very simple pattern: the sine wave. The vital component of a lock-in amplifier is a balanced mixer stage, which is actually a mathematical multiplier. If you insert a sine wave into one port of the mixer (a LO) at exactly the same frequency and phase as the signal you're looking for (inserted into the second input port), the output of the mixer at any point will be the numerical product of your LO and the *signal in question* (SIQ, for our current purposes). This process inherently rejects any signal that isn't perfectly correlated with the LO, while amplifying the daylights out of the correlated signal.

This probably brings up another valid question. What if you don't have a clue what you're looking for in the first place? In this case, the lock-in amplifier is *not* the instrument of choice. However, as we will shortly see, in radio science we often do know exactly what we're looking for, and this is where the lock-in amplifier shines.

We did leave out one small feature of the lock-in amplifier that sets it apart from a "normal" direct conversion receiver: a lock-in amplifier has two receiving channels that operate in quadrature. The in-phase channel is known as the *I* channel; the quadrature channel, which operates 90° out of phase, is called the *Q* channel. I and Q reception and transmission schemes, once rather obscure, have become familiar in the ham radio shack, particularly with modes like QPSK and other sound card digital modes. But let's look at the more primitive analog aspects of the I/Q receiver. With a single-channel I receiver, the output of the mixer stage decreases as the phase between the LO and the SIQ is shifted. At 90° of phase shift, the output is zero. The rate of decrease varies precisely as the sine of the angle between the LO and the SIQ; or, more conventionally notated, the output of the mixer is the product of the LO and the SIQ times the cosine of the angle between them.

This presents a dilemma. We have a receiver, or, more specifically, a mixer, the output of which varies as a function of the amplitude of the SIQ *and* the relative phase of the SIQ. If we see a signal that's varying in output level, how do we know whether it is the phase or the amplitude that's varying? We don't. We've encountered the ambiguity gremlin we mentioned briefly in our section on sensor arrays. In fact, even if the amplitude of the SIQ is absolutely constant, we can't know for sure if it is precisely in phase with our LO, unless we wobble the LO phase around a bit and see that we have maximum output from the mixer. What do we do?

The answer is to build another identical mixer, but with a LO that's precisely 90° out of phase with the I-channel LO. This actually allows us to solve two ambiguities: If the I mixer is precisely in phase with the SIQ, the Q mixer will have *zero* output. If there is any output at all from the Q channel, we know we have a phase error. But beyond that, we can now know *which direction* the phase error is in. This is because a balanced mixer is bipolar; it puts out a positive voltage if the LO and SIQ are in phase, and a negative voltage if they are 180° out of phase.

The modern lock-in amplifier will actually offer two modes of displaying the detected signal, either as an X and Y coordinate, or as a magnitude and a phase angle (θ). The latter is probably a bit more intuitive, and it doesn't require doing the Pythagorean Theorem in your head. Regardless of the phase angle of the incoming signal, the Z axis (magnitude) will be correct.

Most practitioners use a lock-in amplifier in conjunction with an oscilloscope in the X-Y mode, which gives an immediate and satisfying view of what's happening. All self-respecting lock-in amplifiers have a pair of output jacks for doing just this. In addition, as we adapt lock-in methods to higher frequencies and more rapidly varying data, we will *need* to use an oscilloscope in this manner, as our brains and eyes can't follow normal digital readouts fast enough to make any sense of them in real time.

If you do have an opportunity to pick up a lock-in amplifier on the surplus market, don't pass it up. As is the case with any instrumentation with the word "scientific" in front of it, such hardware is seriously overpriced, at least new. However, some good bargains do occasionally show up, and they're still good buys even if the items in question need a bit of TLC and calibration.

Better still, check out some of the homebrew IQ receiver projects that have been presented in recent ham literature as described above. They're great building blocks for many of the HF lock-in methods applied in radio science experiments.

A FEW MODIFICATIONS

As we mentioned, the traditional lock-in method is used primarily for VLF. Most professional lock-in amplifiers have extremely sharp and low frequency cutoff filters available, often performed with sophisticated digital signal processing (DSP) methods. Again, the bandwidth of the basic direct conversion receiver is determined by a low-pass filter after the detector, not by a band-pass filter as in a more conventional receiver. The response of these filters in a typical scientific lock-in amplifier is generally described in terms of *time constant*, rather than cutoff frequency. These time constants can be in the order of tens of minutes, which can put the equivalent cutoff frequency down in the millihertz range!

For many radio science experiments, however, particularly those involving time-of-flight measurements, we will want to look at short pulses, which means one needs to adapt the lock-in to a much higher cutoff frequency. For these applications, it's still useful to refer to the time constant of the filters rather than the cutoff frequency, and we may need something on the order of a 50 μs time constant, or about a 20 kHz equivalent bandwidth. Fortunately, this is very easy to tweak with op-amps, if one is comfortable working with them, and even *easier* to tweak with DSP or even sound card methods.

In **Figure 18.1** we see a direct conversion IQ receiver made in-house for the skymapping project at HIPAS Observatory. The heart of this high-performance receiver is a direct digital synthesis (DDS) module that generates a precise sine and cosine signal we use for the I and Q local

Figure 18.1 — This scientific grade I/Q channel receiver uses direct conversion and a direct digital synthesis chip for the local oscillator. Plenty of "air space" is incorporated in its design, allowing for easy modification, a must for any scientific endeavor. [Thomas Griffith, WL7HP, Silver Impressions, photo]

oscillators. This is seen at the upper right corner. In the middle are the balanced mixers and low-pass filters. There's also a DAQ access block (with the ribbon cable) that we used to interface our computer running National Instruments *LabVIEW* software.

IMPROVING YOUR AVERAGES

There will be instances when the lock-in method just doesn't fit the bill. One of these cases is when you don't have a known frequency with which to compare your SIQ; another case is when you don't really need to know the phase information, but simply want to find something meaningful buried in the noise. This is when the method of signal averaging is highly effective. However, unlike the lock-in amplifier, which operates in real time, any form of signal averaging requires some sort of memory and post processing.

Perhaps the oldest signal averaging method in common use was

Probing Questions

The word "ether" comes down to us from the Greek Aethos, the god of the upper air of heaven. To Aristotle and centuries of theorists after him, ether was the hypothetical substance through which electromagnetic waves travel. Most modern scientists shun the word, but for all its flaws, ether is an extremely useful, if incomplete, model for much of what we observe in radio science, especially when we have to build some practical hardware, specifically a probe to study it. The antenna is the probe with which we extract meaningful information from the ether — and the ether is the sole connection between our probe and just about anything else of radio interest, what we call the *phenomenon under examination* (PUE).

Sometimes there is a straightforward path between the PUE and the probe; sometimes not. A signal from a transmitting antenna wobbles the ether at some frequency, and our receiving antenna responds to the ether wobbling at the same frequency, or there may be an intermediary or two between PUE and probe. Some phenomena cannot be measured directly with an electromagnetic probe; for example, we cannot pick up sound waves with an electromagnetic antenna no matter how loud they are. Sound waves and electromagnetic waves are in two entirely different realms. The fact that sound cannot propagate without a medium while electromagnetic waves can is one major distinction between them. Also, sound waves are longitudinal, while electromagnetic waves are transverse.

So, the wobbling of the air does not create any electromagnetic waves and, therefore, is not detectable by an electromagnetic antenna. Why not, if air is made of atoms, and atoms have charges, and when you wobble a charge you're supposed to get a radio wave? The problem is, when air is wobbling in the form of sound waves, the nuclei and the electrons of said atoms are wobbling at the same time in the same direction, at the same rate of wobble. There is no acceleration of net charge, so anything meaningful from an electromagnetic standpoint is cancelled. Air in its raw state is invisible to electromagnetism. But air is not the only example.

Many exotic particles, including some that physicists are looking for now, have no charge — period. This makes it a particularly interesting challenge to detect a PUE via any electromagnetic methods. And, since they are often few and far between, we can't rely on acoustic phenomena (particles bouncing against like particles) to make themselves known, since there may not be any like particles *anywhere*, well, at least not anywhere in the immediate neighborhood. What's a physicist to do? A particularly useful approach in the physicist's bag of tricks is to employ an intermediary agent between an electromagnetic phenomenon and a non-electromagnetic one. One of the more accessible of these intermediaries is the neutron. A neutron has no charge of its own — hence the name — but it can be knocked around by another high-energy neutral particle, in which case nuclear forces dominate over electromagnetic ones.

Most of us radio folks don't think about neutrons much, except as an intellectual placeholder. However, the neutron can act as a sort of glue to help hold the components of atomic nuclei together. (This is not to be confused with a gluon, which holds the components of the *components* of atomic nuclei together.) When a neutron gets knocked out of a nucleus, it tends to drag a proton or two with it. Once this happens, what is left behind is — you guessed it — an ion. We know that ions *do* respond to electromagnetic waves — or can create them if they are accelerating. Most probes available to us for detecting charge-less particles actually detect a secondary, or even tertiary (perhaps more), effect of a particle collision. Let's look at a few types of probes we may encounter.

If to a man with a hammer, everything looks like a nail, then to the radio amateur, everything looks like an antenna. This is not necessarily a bad thing, but we sometimes need to look at processes that aren't strictly electromagnetic. In our earlier discussion on sensor arrays, we touched on charge detectors, which are basically capacitor plates. In reality, one seldom uses "plain old" capacitor plates for particle detectors, but rather the capacitive properties of various types of semiconductor materials that provide the equivalent of millions of tiny capacitors in a few cubic centimeters. The detector material is a secondary source of charged particles that result from a collision by our real particle of interest. We also talked a bit about magnetic detectors, which aren't strictly antennas, either.

An interesting situation arises at ULF and ELF frequencies as well. If you recall from our discussion of near and far field measurements, every antenna radiates both an electromagnetic field and a near field or inductive field. The actual inductance field does not radiate, but it does attenuate with distance from the antenna elements. When working with ELF radio signals, where the wavelengths can be thousands of kilometers, we also have a "near field" which can extend for tens or hundreds of kilometers. Because of this, there is some debate as to whether ELF submarine communications are really radio *per se*, or whether they are simply magnetic coupling, as one would have in a transformer, albeit with a lot of space between the primary and secondary windings.

Antennas for ELF don't resemble antennas used at "official" radio frequencies. A typical high-performance ELF coil (we don't even call them antennas) is a ferrite rod, three or four feet long, with tens or hundreds of thousands of windings around it. The radiation resistance of such an "antenna" is essentially non-existent at frequencies down in the hundreds of hertz — or less. Likewise, its ability to respond to a radio wave's E field is nil.

The wavelength of a 200 Hz "radio" signal, where typical submarine ELF communications take place, is 1500 kilometers. At both HIPAS Observatory and HAARP, we demonstrated how we could use the Earth's electrojet (the concentrated plasma ring at E-layer levels of the ionosphere, just below the visible Auroral oval) as an effective ELF antenna. Now, as high as this may seem, 90 km or so, it is still only about $\frac{1}{15}$ λ high at 200 Hz, meaning anything on the surface of the Earth (or a few miles down in the ocean) is well within the near field of the "Auroral antenna." It really would be a stretch, then, to consider such communications "radio" in the conventional sense. Incidentally, this particular experiment is the primary reason for the existence of HAARP. Effective ELF submarine communication has been one of the holy grails of ionospheric experimentation for a long time, although a lot of serendipitous science has come out of these efforts as well.

Although radio and sound are in different domains, there are instances when the frequencies of audio and radio overlap. This is the intriguing realm of VLF radio, where a lot of natural radio phenomena occur. One of the most effective probes for VLF radio is simply a high gain audio amplifier — and a few hundred feet of wire. Audio frequencies are generally considered to fall between 20 Hz and about 20,000 Hz, and it's in this range that numerous fascinating ionospheric plasma phenomena occur. Some of the better known of these oddities are whistlers, which are rapidly descending tones, starting at a few kilohertz and ending at a few hundred. They were first observed more than 100 years ago on military field telephones. Whistlers are dispersive ionospheric waves, generally excited by distant lightning storms.

developed by astronomers who were looking for *very* dim objects, especially in the presence of atmospheric disturbances and the like. The method, which was used *long* before the Hubble Telescope, simply involved taking lots of photographs of the same region and then overlaying the plates during development. Random or moving light sources, such as atmospheric reflections, would be canceled out, while more coherent light sources, such as small stars, would be reinforced with each exposure. Obviously, if you were actually looking for a transient object, signal averaging wouldn't work.

For radio science, signal averaging can be performed with any DAQ device and its associated software, or even with a digital storage oscilloscope (the price of which has dropped dramatically in recent years). If you have access to a digital storage oscilloscope, it's really impressive to see what signal averaging can do. You can demonstrate this with a signal generator and a noise source. Or, lacking an "official" noise source, you can just lay one of your scope probes near the signal generator, so all you get is a minuscule amount of leakage into the vertical channel, and crank up the vertical gain until you see a good deal of internal noise. Turn on the signal averaging, and watch the noise floor slowly drop away from the desired signal as if by magic.

Of course, this process takes time. If you take 50 averages, you have to account for 50 times the individual sample time (and a bit more for the processing itself) before your data point is displayed, so there is a definite price to pay in terms of speed of acquisition. Again, this is unsuitable for extremely fast, fleeting phenomena. But for digging continuous signals out of the noise, it can't be beat. You can also simply add the sampled waveforms together, but this causes the desired signal to keep rising, which may be a problem if you need to measure the amplitude, not just see it. Both methods — signal adding and signal averaging — have their rightful place in the research environment.

The reason signal averaging and signal adding work so well is that noise, by its nature, is random. However, the reason it doesn't work perfectly is because, well, noise is not *perfectly* random. Noise itself has recognizable patterns, which, when you think about it, makes as much sense as the existence of "laws of chance." Why should random chance have any laws at all?

This is all part of "Big Question" science, of which radio science is only a small contributor.

Now for Some Real Radio Science

Throughout this book, we've seen that good scientific method is good scientific method, regardless of the particular branch of study. In this chapter, we will present a "soft entry" into actual radio science methods using the principles from our saltwater demonstration described in Chapter 4. We'll tackle what is arguably the most fascinating aspect of radio science accessible to the average ham: ionospheric propagation.

There is still a great deal that we don't know about how radio signals travel from Point A to Point B through the ionosphere. Most radio amateurs who have been operating for any length of time will tell you that the ionosphere is about as predictable as a roulette wheel in an earthquake. However, as we will discover, despite some anomalous behavior, ionospheric propagation bears a high degree of correlation to many other predictable, or at least periodic, events. As we've learned, in order to make any meaningful scientific discoveries, it's crucial to know what's *normal* so we can establish a baseline for our measurements. We can then begin to make meaningful connections between Event Z and Cause Y.

Because Amateur Radio activity is a somewhat sporadic activity for most of us, it is easy to conclude that the ionosphere never behaves as it is "supposed to." Propagation never seems good when we want it to be, such as during a contest or a rare DXpedition. If we were privileged enough (or physically able) to operate as hams 24/7, we would find nearly clockwork variations in ionospheric behavior — at least *most* of the time. Fortunately, thanks to the ready availability of computer control, we can, for all practical purposes, continuously monitor conditions, even while we sleep. In so doing, we stand a very good chance to observe some surprisingly *sane* patterns of behavior of our beloved ionosphere.

To that end, following are a few simple methods of assembling a channel probe, which will allow us to meaningfully measure propagation paths between any fixed set of points over a prolonged period of time.

A NEW MISSION FOR YOUR RECEIVER: LEVEL 1

In this example, we'll use an eclectic combination of traditional and state-of-the-art technology to establish a behavioral baseline for radio propagation. Here are all the ingredients you'll need to do some real radio science.

Demo 19.1a — Materials and Procedure

• A general coverage communications receiver with an analog S meter, any vintage, any technology. I use a 1940 Hallicrafters SX-25 "Super Defiant." Such big, old tube radios, known as boat anchors, have several attractive advantages over newfangled rigs, since it's easier to access certain parts. **Warning**: There are potentially lethal voltages in the bowels of tube-type radios. If you're new to working with them, enlist the help of someone who has experience with these devices.

• A 15 MHz dipole antenna, oriented magnetic north and south. It doesn't have to be very high at all; in fact, there are certain advantages to very low dipoles for ionospheric science. If you don't have room for a full-sized dipole, you can use a loaded dipole, but orient it as close to magnetic north and south as possible, for reasons we'll discuss later. You can check with your local air service station for the precise declination in your neighborhood.

• A DAQ card or USB module.

• A computer with a USB port capable of running DAQ software (the DAQ software will be supplied by the manufacturer of your DAQ card or module). For our purposes, laptops are preferable to desktop computers as they use far less power during continuous operation. The computer doesn't have to be a high-end machine, as long as it will accept the DAQ device and run the software.

• Twisted pair wire

First set up your dipole and connect it to your receiver with a suitable transmission line. Make sure you can obtain a reasonable signal from the 15 MHz WWV station during the day. By "reasonable," we mean the signal is strong enough to move the S meter and does not suffer from too much local electrical noise. If you live in a "bad radio neighborhood," you may have to reorient your antenna to reduce noise, even if you lose the ideal N-S orientation in the process.

Once you're sure you have a decent set of "ears," connect the DAQ device to your radio. First turn off the radio and discharge any exposed connections that might be connected to capacitors, especially the S meter terminals. Again, you might enlist a local Elmer for assistance. Use a twisted pair of wires to connect the S meter to an analog input of your DAQ device, as we did in the saltwater demo. **Caution**: There may be some high voltage on these input terminals, so take measures to avoid touching them inadvertently.

Most DAQ devices can handle up to around 75 V common mode with no problem, as long as the differential voltage (the voltage between the two terminals) is no more than a few volts, which is always the case on even the most decrepit of receivers. However, you will want to actually measure the voltage to ground of the S meter terminals to be sure this doesn't exceed the specifications of the DAQ device. If it does, you can build a simple voltage divider consisting of a 1 MΩ resistor and a 100 kΩ resistor from each terminal of the S meter to reduce the voltage the DAQ device will encounter (**Figure 19.1**). This will decrease the input voltage by about 10:1, but it is still plenty of signal for any DAQ. You may need to adjust one or the other of the 100 kΩ resistors to get a zero reading on the DAQ device when the S meter is zero.

Fire up your DAQ software (you did read all of the instructions that came with it, didn't you?). Now, fire up your receiver, determine that it is operating properly, and tune in WWV. Set up the graphing software of your DAQ device to sample every second or so. Be sure you set it up for continuous data logging; by default some software will quit after some predetermined time. (Again, we cannot overemphasize reading the entire manual.)

Determine that the graph follows the normal variations of the S meter reading. To be on the safe side, you can disconnect the antenna and observe that the graph drops precipitously; if it does, you're good to go. Let your equipment run for a week to collect some data. (Depending on the receiver you use, you may have to occasionally check to make sure it remains tuned to WWV.)

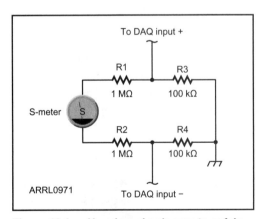

Figure 19.1 — Here is a simple way to safely connect your DAQ device to nearly any receiver's S meter. Typically the S meter will have less than 0.1 V across the terminals, but it may present a high common mode voltage. When in doubt, measure the terminal voltage with respect to ground.

Results

After a few days of data collection, take a peek at your accumulated graph (**Figures 19.2** and **19.3** are sample graphs from data I

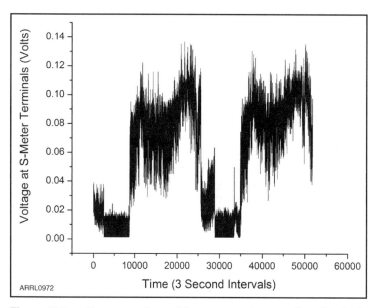

Figure 19.2 — Every once in a while, even a veteran radio scientist is surprised to get such great results. Notice the nearly identical signal strength profiles for two subsequent days, beginning at midnight, September 21, 2012, and ending at midnight, September 23, 2012.

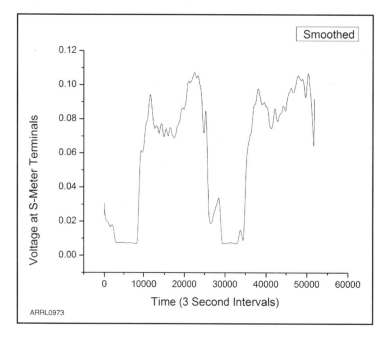

collected). If everything is working, you should see an almost clockwork-like variation of signal strength plotted, with the maximum occurring near high noon and the minimums occurring around midnight. The positive "humps" of the curve will be nearly sinusoidal, while the bottom of the cycle may be pretty flat. The overall appearance should resemble something like the output of a half-wave rectifier.

This regular variation in signal strength or *diurnal variation* (**Figure 19.4** shows another sample graph) is a result of the change of the ionization level with respect to the angle of the Sun. You will most likely see a lot of small zigs and zags on the curve in addition to the sine pattern due to local disturbances in the ionosphere.

Figure 19.3 — Here is the same data we saw in Figure 19.2, but with 1000-point averaging, thus filtering out all the normal short-term signal strength variation one expects over a long path, in this case, Boulder, Colorado, to Fairbanks, Alaska.

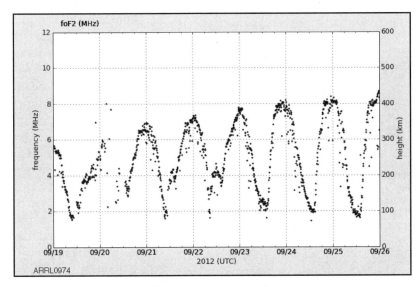

Figure 19.4 — Here is the HAARP ionosonde plot of diurnal (daily) variation of critical frequency overlapping the same period of time as the S meter trial (Demo 19.1a). Notice the beautiful sinusoidal variation of critical frequency with respect to daylight hours, which also correlates with the WWV S meter data. This is great "baseline" science — gathering reliable data under "normal" conditions, which can be used as a reference for experimentation.

However, if we were to filter those out with some mathematical smoothing, we will see the more meaningful long-term pattern. We have now established one of the most important correlations in all of radio science: the connection between ionization and solar input.

Possessing a properly functioning gift of suspicion as we do, we can't just leave it at that: We want to verify our suspected correlation by actually measuring local sunlight levels to determine if the correlation is real or just our imagination. Let's now add that element to our current experiment setup.

Demo 19.1b — Expanded Materials and Procedure

- Photoresistor
- 9-volt battery
- 1 foot of ½ inch PVC tubing

One of the cheapest sunlight detectors is a photoresistor and a 9 V radio battery (**Figure 19.5**). Since our DAQ device has all these extra input channels, why don't we simultaneously record local sunlight levels on Channel

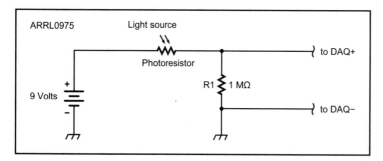

Figure 19.5 — Here is a "cheap and dirty" method for obtaining outdoor light levels to test the correlation of mid-path MUF with local daylight. The photoresistor and power source may be replaced with a photocell (solar cell) if desired. In either case, rigorous scientific technique dictates that the solar detector face the true north horizon.

2? Although you can just hang your photoresistor out a random window and get a pretty good reading of daylight, standard practice has the photodetector oriented true north. To exclude manmade light sources, it's a good idea to place the detector in one end of a small tube, say a foot long piece of ½ inch PVC, and aim the opposite end of the tube at the northern horizon. Again, use a twisted pair of wires to connect your photoresistor to your DAQ device in order to reduce electrical noise pickup.

Since resistance in a photoresistor decreases with the amount of sunlight, you will want to apply the 9 V battery's positive terminal to the "hot" side of the photoresistor, and apply the "low end" of the photoresistor to the positive DAQ input lead. The negative lead of the DAQ input will go to the battery's negative terminal. A 9 V battery will last for months (possibly much longer) of continual operation in this application as the input impedance of the DAQ device is extremely high, taking almost no current.

Now, you can rerun your experiment for another week. Your DAQ software will most likely give you the option of creating two graphs, one above the other, or a single graph with two different colored plots, one for each channel. Both of these viewpoints have their advantages, and I use them both extensively.

Results

After running your "new and improved" experiment for a while, you may notice that the local sunlight peak may lead or lag the diurnal signal strength peak readings by an hour or two — or more. What do you see? (To help get your analytic brain cells firing, we offer several test questions on this and other findings at the end of this chapter.)

THE NEXT STEP: LEVEL 2

I don't think I've ever met a ham who had just one receiver or antenna; most of us have multiples of most of our equipment. Hopefully that describes you, too, because the next experiment will require two communications receivers and two antennas. Ideally, the receivers will be identical models, but it's not necessary; we can fairly reliably calibrate out the differences between them. The second antenna, however, *does* have to be identical to the first one, except at right angles — a "magnetic east-west" 15 MHz dipole. We enclose the term in quotes because, obviously, there's no such thing as a magnetic east or west pole; we simply mean that the antenna is perpendicular to the north-south magnetic pole antenna.

(Lacking a second receiver, we can achieve comparable results by using a coaxial relay to periodically switch between the N-S antenna and E-W antennas.)

Demo 19.2 — Expanded Materials and Procedure

- A second general coverage communications, OR a coaxial relay
- A second 15 MHz dipole antenna, oriented "magnetic east and west"
- 4066 analog switch chip (recommended with coaxial relay setup)

By adding a second setup — whether it's the two complete receivers and antennas or one receiver with coaxial relay to two antennas — we can now acquire even more useful baseline information about the ionosphere. If you use the switching method and a single signal-strength plot, you'll need to have a readily identifiable time stamp to keep track of which antenna is being plotted at which time. This can be a bit awkward so it's better to use another electronic switch to alternately route the S meter voltage to two different analog inputs of the DAQ device; you can then isolate the two antennas into two graphs. A 4066 analog switch chip is ideal for this (**Figure 19.6**).

With your two receivers, now re-run your experiment for a week and see if you note any interesting trends in the two S meter plots.

Results

It is a well-known phenomenon that very low dipoles over practical ground are nearly non-directional. You can generally verify this during the daytime by switching between your two dipoles while listening to WWV. However, after running the experiment for a few days, you may find that the long-term diurnal variations on the E-W receiver do not

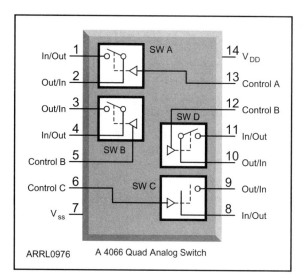

Figure 19.6 — A low-cost analog switch, such as a 4066, can make the routing of DAQ signals fast and simple. This device can be driven with one of the digital output lines of a typical low-cost DAQ device. This particular chip has four bi-directional analog switches, which add no distortion to the analog signals as long as they are within specifications.

precisely follow those of the N-S receiver. In addition, you may find there is a profound difference between the two receivers when it comes to short term fading. What did your data indicate? Can you put forth an explanation?

DELVING DEEPER STILL FOR PHASE: LEVEL 3

Although we can determine (or at least *infer*) a great deal of information about the ionosphere by looking at signal strength alone, there are a lot of questions this leaves unanswered. In addition to measuring signal strength, to do complete radio science we need to look at something generally ignored by radio amateurs: the *phase* of incoming radio signals. At this level, we aren't as concerned about *absolute phase*, which is somewhat hard to define anyway, but rather the *relative phasing* between the E-W and the N-S receivers. Fortunately, one of the oldest and simplest receiver technologies — the direct conversion receiver — will give us the phase information we need.

We touched on the important principle of quadrature earlier in this book, but we need to revisit it now. To fully and unambiguously measure phase angle, we need two receivers operated in quadrature. A balanced mixer by itself (the core of the direct conversion receiver) will give us an output level that is proportional to the cosine of the phase angle between the incoming RF signal and the LO. This is fine, as long as the signal strength of the incoming signal is constant or known. However, if the incoming signal strength is unknown, or varying, our direct conversion receiver can't tell us if it's the phase or the amplitude that's changing.

If we have a means of measuring signal strength that is independent of the phase angle, such as our receiver's S meter, we can resolve this ambiguity. This approach is awkward to implement, however; it's far easier and practical to build two identical direct conversion receivers whose LOs are in phase quadrature (90° phase difference), the output of each of which is then applied to a channel of our DAQ device. The two channels of voltage emerging from a quadrature receiver are known as the I and Q signals (in phase and quadrature, respectively).

HOMEWORK — RECYCLING TECHNOLOGY FOR REAL SCIENCE

Rather than reinvent the wheel, for your last step in our "Real Science" demonstration we're going to give you a homework assignment. First get your hands on a copy of the excellent March 1999 *QST* article, "A Binaural IQ Receiver" by Rick Campbell, KK7B (ARRL members can do a *QST* archive search as **www.arrl.org/qst**), give it a considered read, and then go *build* the quadrature receiver Campbell outlines. Try as I might, I can't imagine ever coming up with a better way of producing

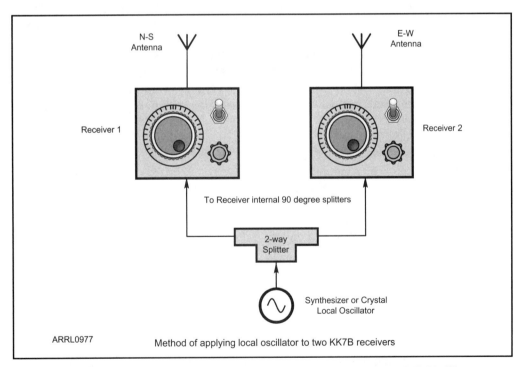

Figure 19.7 — The only way to improve on a great receiver design, such as KK7B's I/Q "binaural" receiver (see text), is to make two of them — and use the same local oscillator for both. Any LO scheme will work for this tweak, but it should be as stable as possible.

this basic circuit. Although Campbell's reason for building this receiver was quite different from ours, his design is perfect for this experiment — with one small twist. You guessed it: we'll need to build *two* of them, one for the E-W signal and one for the N-S signal. However, we can save a bit of time and expense, because we can use the same LO for both receivers; a simple signal splitter will do the trick to route the LO to all four mixers. (**Figure 19.7**).

Demo 19.3 — Expanded Materials and Procedure

- 2 identical quadrature receivers built per KK7B design (see March 1999 *QST*)
- Signal splitter
- Additional DAQ cards or modules (recommended)

Proceeding with the assumption that you've bitten the bullet and built two identical quadrature receivers, we'll take the last step and turn our setup into a first-class radio science station that will be capable of doing just about any other radio science experiment your curiosity leads you to.

Having replaced radios of the earlier demos with these quadrature receivers, you will now have to dedicate four DAQ analog input ports to the task. You may have to sacrifice the photodetector if you have only four DAQ channels available on your DAQ device, or get a second DAQ device, or a third, or a fourth.

Let's run our experiment again for a week or so and examine the new results. If you look at the two plots of a given antenna, say the N-S dipole, you'll see that they seldom match. If you look at the sums of the plots at any particular time, you'll see the signal strength at that moment (your software can probably do this task for you). More than likely, you will observe some radical flopping between the two channels' data of a given antenna.

If you now find that you have more questions than answers, don't be alarmed — it means you have simply discovered some more interesting radio science.

STANDING THE TEST OF RADIO SCIENCE

One of the premises of *Radio Science for the Radio Amateur* is that we live in a fairly coherent universe, with a lot of cross-correlation between basic physical principles. The fact that we can use our familiar Amateur Radio equipment and basic technical skills to explore so much of other scientific specialties is nothing short of astonishing. We hope that this book may inspire you to take up this journey of discovery and see where it leads you — perhaps to the ionosphere and beyond.

Now, as promised, following are some radio science "test questions" to challenge your inquisitive mind. Answers can be found in **Appendix I**.

Q1. After running Demo 19.1b, you may notice that the local sunlight peak may lead or lag the diurnal signal strength peak readings by an hour or two, or more. Why might this be?

Q2. After running Demo 19.2, how do you explain the variations between the N-S and the E-W signal strengths?

Q3. What information about the ionosphere can we obtain by looking at the separate I and Q plots of a given antenna?

Q4. How can we use our four channels of data to determine the polarization of the incoming radio signal?

Q5. How can we determine the direction of arrival using our four channels of information?

Q6. Can we unambiguously determine both the angle of arrival and the polarization using our four-channel setup? Why or why not?

Appendix I
Test Questions and Answers

Here are answers to the questions we posed in Chapter 19:

Q1. After running Demo 19.1b for a while, you may notice that the local sunlight peak may lead or lag the diurnal signal strength peak readings by an hour or two, or more. Why might this be?

A1. In general, it's the ionization level at a location midway between the transmitter and the receiver that determines the propagation characteristics. Unless you happen to live right in Boulder, Colorado, where WWV is, the Sun won't be at its peak altitude at mid-path at your local high noon. This is, of course, a gross oversimplification, since in many cases, there are multiple hops between transmitter and receiver. On the average, however, you will find that the strongest signal occurs at around high noon of the mid-path.

Q2. After running Demo 19.2, how do you explain the variations between the N-S and the E-W signal strengths?

A2. The ionospheric plasma is magnetized by the Earth's magnetic field, which causes the ionosphere to have very different properties moving along the magnetic field lines as opposed to across them.

Q3. What information about the ionosphere can we obtain by looking at the separate I and Q plots of a given antenna?

A3. By acquiring both I and Q signals, we can easily measure the movement of the ionosphere. Movement of the ionosphere creates Doppler shift, which is an important ingredient of ionospheric science. If you have an X-Y oscilloscope, you can insert the I channel into the X input and the Q channel into the Y input and observe the Doppler shift by watching the rotation of the oscilloscope display.

Q4. How can we use our four channels of data to determine the polarization of the incoming radio signal?

A4. Having four channels of information, an I and Q channel on each of two cross-polarized dipoles, allows you to determine the polarization of signals arriving at high angles. You can tell if the polarization is primarily E-W, N-S, or circularly polarized.

Q5. How can we determine the direction of arrival using our four channels of information?

A5. This is a bit of a trick question. We need to carefully determine what we mean by direction of arrival. For most hams, direction of arrival means direction in azimuth; that is, what compass heading are you using. For a lot of radio science, angle of arrival in the elevation plane is at least as important as azimuth angle, if not more so. We will find that our low height crossed dipoles are nearly useless in themselves for determining azimuth arrival angles, but are very useful for determining vertical angle of arrival, especially if the arriving signal is circularly polarized, which applies to nearly all ionospheric signals we will encounter.

Q6. Can we unambiguously determine both the angle of arrival and the polarization using our four-channel setup? Why or why not?

A6. No. This is one of the cruel facts of life when it comes to electromagnetism. To unambiguously determine the direction of arrival, you need to know the polarization; to unambiguously determine the polarization, you need to know the direction of arrival. To resolve this, you need two antennas, or arrays of antennas. Fortunately, for a lot of ionospheric studies, we already have *a priori* knowledge of the polarization; it will be either right-hand circular or left-hand circular. But truly rigorous science does require us to independently measure the polarization and the direction of arrival. Again, this cannot be done with a single antenna.

Appendix II
Alphabetical Formulary

Following is a short list of useful equations, relationships, and identities for the amateur experimenter, sorted alphabetically by their commonly known names. This is by no means a comprehensive listing of all the mathematical formulas one might use in radio science. We have no intention of competing with *The Handbook of Chemistry and Physics* or any other such ponderous tome, but it should prove a helpful reference.

Apparent Power: $P = I \times E$

Appleton-Hartree Dispersion Relation:

$$n^2 = 1 - \frac{X(1-X)}{(1-X) - \frac{1}{2}Y_T^2 \pm \left[\frac{1}{4}Y_T^4 + (1-X)^2 Y_L^2\right]^{1/2}}$$

Bandwidth (half power): $B = f / Q$

Capacitive Reactance: $X_C = \dfrac{1}{2\pi f c}$

Complex Impedance: $Z = \sqrt{x^2 + r^2}$

Conductance: $G = 1 / R$

Decibel: $dB = 10 \log_{10}\left(\dfrac{P_2}{P_1}\right)$

Decibel (with equal input and output impedance):

$dB = 20 \log_{10}\left(\dfrac{V_2}{V_1}\right)$

Dissipation Factor: $D = 1 / Q$

Energy: $E = P \times T$

Frequency: $f = 1 / \text{period}$

Impedance: See Complex Impedance

Inductive Reactance: $X_L = 2\pi f L$

Inductance of solenoid (single layer, air core) in microhenrys:

$$L = \frac{N^2 D^2}{18D + 40H}$$

where
- N = number of turns
- D = diameter of coil in inches
- H = length of the coil in inches

(Note: This formula was recently revised to include the radius of the wire, but for practical purposes, this can be ignored.)

Kirchhoff Current Law (KCL): Total current entering a node must equal the current exiting the node.

Kirchhoff Voltage Law (KVL): Sum of the voltage drops around a closed loop must equal 0.

Law of Optimum Radio Performance (LORP): See *TCC*.

Mho: Old name for siemens, the unit of *conductance*. Mho is actually preferable to siemens for most radio amateurs, because it's easy to remember that with Mo' mho you have mo' current flow.

Ohm's Law: $E = I \times R$

Ohm's Law for Reactive Circuits: $E = I \times R$

Ohm's Law for Complex Circuits: $E = I \times Z$

Power (continuous): $P = I \times E$

Power (intermittent): $P = E / T$
where E = energy and T = time.

Q: $Q = \dfrac{X_L}{R}$

(Note: While Q is sometimes defined simply as X/R, this can be misleading in the case of a resonant circuit where the *net* reactance is 0. Since a *practical* inductor in a complex circuit is most frequently the limiting factor for circuit Q, it is the inductive reactance that is most often the numerator.)

Reactance (net): $X_t = X_L - X_C$

(Note: Since the net reactance term is squared in calculating complex impedance (Z), this number will always be positive, so you don't need to worry about the order, unless you need the phase angle.)

Resistance: $R = E / I$

Resistances in Series: Add individual resistances

Resistances in Parallel: Convert individual resistance values to conductance values and add conductance values to obtain total conductance. Take the reciprocal of total conductance to obtain total resistance.

Resonant Frequency: $f = \dfrac{1}{2\pi\sqrt{LC}}$

Solenoid: See Inductance

TCC (Total Copper Content): A useful measure of the effectiveness of any ham radio station.

Transformers: Voltage ratio is proportional to the turns ratio.

Impedance ratio is proportional to the *square* of the turns ratio.

Transmission Lines (input impedance):

$$Z_{in}(l) = Z_0 \frac{Z_L + jZ_0 \tan(\beta l)}{Z_0 + jZ_L \tan(\beta l)}$$

where
  l = the distance from the load
  β = the wave number.

(Note: Now you see why I like the Smith Chart!)

Transmission Line Q Section: $Z_{in} = \dfrac{(Z_0)^2}{Z_{LOAD}}$

True Power: $P_{TRUE} = I \times E \cos(\theta)$

where
  θ = the phase angle between voltage and current

Volt-Amperes-Reactive (VAR): Simple product of voltage and current in a reactive circuit, neglecting phase angle.

Watt: $I \times E$, E^2 / R, $I^2 \times R$

Z (complex impedance, long form): $Z = \sqrt{(X_L - X_C)^2 + R^2}$

SPECIAL BONUS

Many of my electronics students wonder how we derive the formula for resonant frequency:

$$f = \frac{1}{2\pi\sqrt{LC}}$$

Doing the derivation of this isn't difficult, and it reveals a lot about the physical concept. First, by definition, a resonant circuit is one in which $X_L = X_C$. So we set up the original equality with the substitute values:

$$2\pi fL = \frac{1}{2\pi fC}$$

Then all we need to do is solve for f. First collect all the terms on the left:

$$2\pi fL 2\pi fC = 1$$

Next, we isolate ff on the left side:

$$ff = \frac{1}{2\pi 2\pi LC}$$

We now make the formula a little more "politically correct":

$$f^2 = \frac{1}{4\pi^2 LC}$$

Now, all we have to do is take the square roots of both sides of the equation, and voila:

$$f = \frac{1}{2\pi\sqrt{LC}}$$

What could be simpler? What a great way to end a book!

Please use this form to give us your comments on this book and what you'd like to see in future editions, or e-mail us at **pubsfdbk@arrl.org** (publications feedback). If you use e-mail, please include your name, call, e-mail address and the book title, edition, and printing in the body of your message. Also, please indicate whether or not you are an ARRL member.

Where did you purchase this book?
☐ From ARRL directly ☐ From an ARRL dealer

Is there a dealer who carries ARRL publications within:
☐ 5 miles ☐ 15 miles ☐ 30 miles of your location? ☐ Not sure

License class:
☐ Novice ☐ Technician ☐ Technician Plus ☐ General ☐ Advanced ☐ Amateur Extra

---

Name _____ ARRL member? ☐ Yes ☐ No
_____ Call Sign _____
Daytime Phone ( ) _____ Age _____
Address _____ E-mail _____
City, State/Province, ZIP/Postal Code _____
If licensed, how long? _____
Other hobbies _____

For ARRL use only	RSRA
Edition	1 2 3 4 5 6 7 8 9 10 11 12
Printing	1 2 3 4 5 6 7 8 9 10 11 12

Occupation _____

From _____

| Please affix postage. Post Office will not deliver without postage. |

EDITOR, RADIO SCIENCE FOR THE RADIO AMATEUR
AMERICAN RADIO RELAY LEAGUE
225 MAIN STREET
NEWINGTON CT 06111-1494

— — — — — — — — — please fold and tape — — — — — — — — — —